Environmental Science, Engineering and Technology

Gulf Oil Spill of 2010: Liability and Damage Issues

ENVIRONMENTAL SCIENCE, ENGINEERING AND TECHNOLOGY

Additional books in this series can be found on Nova's website under the Series tab.

Additional E-books in this series can be found on Nova's website under the E-books tab.

POLLUTION SCIENCE, TECHNOLOGY AND ABATEMENT

Additional books in this series can be found on Nova's website under the Series tab.

Additional E-books in this series can be found on Nova's website under the E-books tab.

Environmental Science, Engineering and Technology

GULF OIL SPILL OF 2010: LIABILITY AND DAMAGE ISSUES

Christine R. Walsh
and
James P. Duncan
Editors

Nova Science Publishers, Inc.
New York

Copyright © 2012 by Nova Science Publishers, Inc.

All rights reserved. No part of this book may be reproduced, stored in a retrieval system or transmitted in any form or by any means: electronic, electrostatic, magnetic, tape, mechanical photocopying, recording or otherwise without the written permission of the Publisher.

For permission to use material from this book please contact us:
Telephone 631-231-7269; Fax 631-231-8175
Web Site: http://www.novapublishers.com

NOTICE TO THE READER

The Publisher has taken reasonable care in the preparation of this book, but makes no expressed or implied warranty of any kind and assumes no responsibility for any errors or omissions. No liability is assumed for incidental or consequential damages in connection with or arising out of information contained in this book. The Publisher shall not be liable for any special, consequential, or exemplary damages resulting, in whole or in part, from the readers' use of, or reliance upon, this material. Any parts of this book based on government reports are so indicated and copyright is claimed for those parts to the extent applicable to compilations of such works.

Independent verification should be sought for any data, advice or recommendations contained in this book. In addition, no responsibility is assumed by the publisher for any injury and/or damage to persons or property arising from any methods, products, instructions, ideas or otherwise contained in this publication.

This publication is designed to provide accurate and authoritative information with regard to the subject matter covered herein. It is sold with the clear understanding that the Publisher is not engaged in rendering legal or any other professional services. If legal or any other expert assistance is required, the services of a competent person should be sought. FROM A DECLARATION OF PARTICIPANTS JOINTLY ADOPTED BY A COMMITTEE OF THE AMERICAN BAR ASSOCIATION AND A COMMITTEE OF PUBLISHERS.

Additional color graphics may be available in the e-book version of this book.

Library of Congress Cataloging-in-Publication Data

Gulf oil spill of 2010 : liability and damage issues / editors, Christine R. Walsh and James P. Duncan.
 p. cm.
 Includes index.
 ISBN 978-1-61324-729-7 (hardcover)
 1. Liability for oil pollution damages--Gulf Coast (U.S.) 2. Oil spills--Law and legislation--Gulf Coast (U.S.) 3. BP Deepwater Horizon Explosion and Oil Spill, 2010. 4. Liability for oil pollution damages--United States. 5. Oil spills--Law and legislation--United States. 6. Marine pollution--Law and legislation--United States. I. Walsh, Christine R. II. Duncan, James P., 1971- III. United States. Oil Pollution Act of 1990. Selections.
 KF1299.W38G85 2011
 344.7304'6332--dc23
                        2011017305

Published by Nova Science Publishers, Inc. † New York

CONTENTS

Preface		vii
Chapter 1	Liability and Compensation Issues Raised by the 2010 Gulf Oil Spill *Jonathan L. Ramseur*	1
Chapter 2	Oil Pollution Act of 1990 (OPA): Liability of Responsible Parties *James E. Nichols*	35
Chapter 3	Liability and Compensation Requirements under the Oil Pollution Act *National Commission on the BP Deepwater Horizon Oil Spill and Offshore Drilling*	61
Chapter 4	The 2010 Oil Spill: Natural Resource Damage Assessment under the Oil Pollution Act *Kristina Alexander*	75
Chapter 5	Natural Resource Damage Assessment: Evolution, Current Practice, and Preliminary Findings Related to the *Deepwater Horizon* Oil Spill *National Commission on the BP Deepwater Horizon Oil Spill and Offshore Drilling*	95

Chapter 6	Cost of Major Spills May Impact Viability of Oil Spill Liability Trust Fund *United States Government Accountability Office Statement of Susan A. Fleming, Director Physical Infrastructure*	**117**
Chapter 7	Unlawful Discharges of Oil: Legal Authorities for Civil and Criminal Enforcement and Damage Recovery *National Commission on the BP Deepwater Horizon Oil Spill and Offshore Drilling*	**143**
Chapter 8	The 2010 Oil Spill: Criminal Liability under Wildlife Laws *Kristina Alexander*	**159**
Index		**173**

PREFACE

The 2010 Deepwater Horizon incident produced the largest oil spill that has occurred in U.S. waters, releasing more than 200 million gallons into the Gulf of Mexico. BP has estimated the combined oil spill costs, including cleanup activities, natural resource and economic damages, potential Clean Water Act (CWA) penalties, to be approximately $41 billion. This book examines the many issues raised by the spill for policymakers, including the ability of the existing oil spill liability and compensation framework to respond to a catastrophic spill. The framework determines who is responsible for paying for oil spill cleanup costs and the economic and natural resource damages from an oil spill; how these costs and damages are defined and the degree to which, and conditions in which, the costs and damages are limited and/or shared by other parties, including general taxpayers.

Chapter 1- The 2010 *Deepwater Horizon* incident produced the largest oil spill that has occurred in U.S. waters, releasing more than 200 million gallons into the Gulf of Mexico. BP has estimated the combined oil spill costs—cleanup activities, natural resource and economic damages, potential Clean Water Act (CWA) penalties, and other obligations—will be approximately $41 billion.

The *Deepwater Horizon* oil spill raised many issues for policymakers, including the ability of the existing oil spill liability and compensation framework to respond to a catastrophic spill. This framework determines (1) who is responsible for paying for oil spill cleanup costs and the economic and natural resource damages from an oil spill; (2) how these costs and damages are defined (i.e., what is covered?); and (3) the degree to which, and conditions in which, the costs and damages are limited and/or shared by other parties, including general taxpayers.

Chapter 2- The Oil Pollution Act of 1990 (OPA) establishes a framework that addresses the liability of responsible parties in connection with the discharge of oil into the navigable waters of the United States, adjoining shorelines, or the exclusive economic zone. Among other provisions, OPA limits certain liabilities of a responsible party in connection with discharges of oil into such areas. The liability limitations established by OPA are currently the subject of significant congressional interest in the wake of the Deepwater Horizon oil spill in the Gulf of Mexico.

Chapter 3- In November 2010, BP estimated that its total costs from the *Deepwater Horizon* spill, including the clean-up, penalties and damages, will total nearly forty billion dollars.[1] BP has the resources to pay this enormous sum. Those who have suffered individual damages from the spill and those who wish to see the Gulf's natural resources restored are fortunate that BP, rather than a smaller oil and gas company, was responsible for the spill. However, the fact that BP is able to provide full monetary compensation for damages that it causes is no more than a fortuity, not a product of regulatory design. If a company with less financial means had caused the spill,[2] the company would likely have declared bankruptcy long before paying anything close to the damages caused.[3]

Chapter 4- The 2010 *Deepwater Horizon* oil spill leaked an estimated 4.1 million barrels of oil into the Gulf of Mexico, damaging the waters, shores, and marshes, and the fish and wildlife that live there. The Oil Pollution Act (OPA) establishes a process for assessing the damages to those natural resources and assigning responsibility for restoration to the parties responsible. BP was named the responsible party for the spill. The Natural Resources Damage Assessment (NRDA) process allows Trustees of affected states and the federal government (and Indian tribes and foreign governments, if applicable) to determine the levels of harm and the appropriate remedies.

Chapter 5- Staff Working Papers are written by the staff of the BP Deepwater Horizon Oil Spill Commission for the use of the members of the Commission. They do not necessarily reflect the views of the Commission as a whole or any of its members. In addition, they may be based in part on confidential interviews with government and non-government personnel.

Six months after the oil has stopped flowing from BP's damaged Macondo well, the amount of environmental harm caused by the spill is uncertain, as is the adequacy of existing legal, regulatory, and policy mechanisms to ensure that restoration needed to redress the damage will be fully implemented by government and paid for by responsible parties. This background paper describes the process that was established under the Oil Pollution Act of 1990 for assessing natural resource damages caused by the spill and restoring damaged resources to their pre-spill condition. Known as Natural Resource Damage Assessment (NRDA), this process is still in the early phases of being applied to the BP spill and conclusions about its efficacy or success in this instance will be impossible to draw for a number of years, possibly decades. This background paper describes the history and purpose of the NRDA, reviews the main steps in the NRDA process, and reports on the status of current damage assessment efforts in the Gulf.

Chapter 6- On April 20, 2010, an explosion at the mobile offshore drilling unit *Deepwater Horizon* resulted in a massive oil spill in the Gulf of Mexico. The spill's total cost is unknown, but may result in considerable costs to the private sector, as well as federal, state, and local governments. The Oil Pollution Act of 1990 (OPA) set up a system that places the liability— up to specified limits—on the responsible party. The Oil Spill Liability Trust Fund (Fund), administered by the Coast Guard, pays for costs not paid for by the responsible party.

Chapter 7- The purpose of this staff working paper is to provide an overview of the sources and uses of penalties and other funds recovered as a result of unpermitted discharges of oil.[1] There are a number of provisions in federal environmental statutes that authorize the federal and/or state governments to seek fines and penalties for violations, recover clean up and removal costs, and secure funds to restore natural resources.

Chapter 8- The United States has laws that make it illegal to harm protected wildlife. Those laws could be used to prosecute those who caused the 2010 oil spill. Perhaps the most famous of these laws is the Endangered Species Act (ESA), which provides for both criminal and civil penalties for acts that harm species listed under the act. The Marine Mammal Protection Act (MMPA) also provides for civil and criminal punishment when an action takes a marine mammal. The Migratory Bird Treaty Act (MBTA) makes it a crime to kill migratory birds.

In: Gulf Oil Spill of 2010...
Editors: C. R. Walsh, J. P. Duncan

ISBN: 978-1-61324-729-7
© 2012 Nova Science Publishers, Inc.

Chapter 1

LIABILITY AND COMPENSATION ISSUES RAISED BY THE 2010 GULF OIL SPILL[*]

Jonathan L. Ramseur

SUMMARY

The 2010 *Deepwater Horizon* incident produced the largest oil spill that has occurred in U.S. waters, releasing more than 200 million gallons into the Gulf of Mexico. BP has estimated the combined oil spill costs—cleanup activities, natural resource and economic damages, potential Clean Water Act (CWA) penalties, and other obligations—will be approximately $41 billion.

The *Deepwater Horizon* oil spill raised many issues for policymakers, including the ability of the existing oil spill liability and compensation framework to respond to a catastrophic spill. This framework determines (1) who is responsible for paying for oil spill cleanup costs and the economic and natural resource damages from an oil spill; (2) how these costs and damages are defined (i.e., what is covered?); and (3) the degree to which, and conditions in which, the costs and damages are limited and/or shared by other parties, including general taxpayers.

The existing framework includes a combination of elements that distribute the costs of an oil spill between the responsible party (or

---

[*] This is an edited, reformatted and augmented version of a Congressional Research Service publication, CRS Report for Congress R41679, from www.crs.gov, dated March 11, 2011.

parties) and the Oil Spill Liability Trust Fund (OSLTF), which is largely financed through a per-barrel tax on domestic and imported oil. Responsible parties are liable up to their liability caps (if applicable); the trust fund covers costs above liability limits up to a per-incident cap of $1 billion.

Policymakers may want to consider the magnitude of the *Deepwater Horizon* incident and the liability and compensation issues raised under a scenario in which BP had refused to finance response activities or establish a claims process to comply with the relevant OPA provisions. BP has either directly funded oil spill response operations or reimbursed the federal government for actions taken by various agencies. BP has paid damage claims well above its liability limit and outside the scope of its liable damages.

Although evidence indicates that the levels of current framework (liability limits and OSLTF) may be sufficient to address the more common mix of spills that have historically occurred, the current combination of liability limits and $1 billion per-incident OSLTF cap is not sufficient to withstand a spill with damages/costs that exceed a responsible party's liability limit by $1 billion. Even if the per-incident cap were increased, the current (and projected) level of funds in the OSLTF may not be sufficient to address costs from a catastrophic spill.

The options available to address these issues depend upon the overall objective of Congress. One objective—which has been expressed by many in and outside Congress—is to provide full restoration and timely compensation (i.e., through channels other than litigation) for the impacts from the spill, without directly burdening the general taxpayers. If this is the objective, Congress may consider some combination of (1) increasing the offshore facility liability limit and corresponding financial responsibility demonstration; (2) increasing the OSLTF per-incident cap; or (3) increasing the level of funds available in the OSLTF. In addition, policymakers may want to consider an industry-financed fund, akin to the nuclear power industry's fund, that could supplement or potentially replace the current system.

Another objective might be to maintain the existing system, which may be sufficient to address all but the most extreme scenarios. Catastrophic spills in U.S. waters have historically been rare. Some may argue that establishing a system that can withstand a catastrophic event would impose costs and yield consequences that would not justify the (expected) ability to address a catastrophic event.

INTRODUCTION

On April 20, 2010, the *Deepwater Horizon* oil drilling rig was nearing completion of BP's deepwater oil well when an uncontained release of hydrocarbons (oil and natural gas) caused explosions and fire, resulting in 11 crew member fatalities. The incident produced the largest oil spill that has occurred in U.S. waters, releasing more than 200 million gallons over approximately 84 days.[1] Although several companies were and are involved (to varying degrees) with the *Deepwater Horizon* incident, BP was (and continues to be) the most prominent private party in oil spill response and compensation activities. Thus, for the purpose of this report, BP is discussed as if it is the sole responsible party—a key term in the existing liability and compensation framework.[2]

The United States has not encountered a spill comparable to the 2010 Gulf spill since the 1989 *Exxon Valdez* in Prince William Sound, Alaska. The *Exxon Valdez* spill tallied approximately $2 billion in cleanup costs and $1 billion in natural resource damages in 1990 dollars. These combined figures equate to approximately $5 billion in today's dollars and do not include the wider array of claims for which responsible parties are now liable.[3]

The total costs of the 2010 Gulf spill are projected to dwarf those of the *Exxon Valdez*. In its 2010 financial statement, BP estimated the combined oil spill costs—cleanup, natural resource and economic damages, potential Clean Water Act (CWA) penalties, and other obligations—will be approximately $41 billion. This estimate includes payments made to date as well as projected future payments, such as claims. However, BP acknowledges the difficulty in estimating some costs and does not include these costs in its projection.[4] Therefore, this estimate is subject to considerable uncertainty.

The incident received considerable attention in 2010,[5] highlighting multiple policy matters regarding oil spills and their aftermath. An issue that has generated (and to some degree continues to generate) particular interest is the oil spill liability and compensation framework.[6] This framework, which is grounded in federal statute and regulations, determines the following:

1) who is responsible for paying for oil spill cleanup costs;
2) who is responsible for paying for economic and natural resource damages associated with an oil spill;
3) how these costs and damages are defined (i.e., what is covered); and
4) the degree to which (or conditions in which) the costs and damages are limited and/or shared by other parties, including general taxpayers.

The first section of this report provides an overview of the existing liability and compensation framework. The second section highlights many of the liability and compensation issues raised by the *Deepwater Horizon* event. The third section discusses options for policymakers to adjust, amend, or supplement the current framework.

EXISTING LIABILITY AND COMPENSATION FRAMEWORK

President George H. W. Bush signed into law the Oil Pollution Act of 1990 (OPA)[7] on August 18, 1990, consolidating existing federal oil spill laws, expanding authorities within the CWA, and creating new provisions regarding oil spill liability and compensation.[8]

The OPA liability and compensation framework includes a combination of elements that distribute the costs of an oil spill between the responsible party (or parties) and a trust fund, which is largely financed through a per-barrel tax on domestic and imported oil. Responsible parties are liable up to their liability caps (if applicable); the Oil Spill Liability Trust Fund covers costs above liability limits up to a per-incident cap of $1 billion. These elements are discussed in some detail below.

Responsible Party

A critical term and concept in the OPA liability and compensation framework is the responsible party. The liability provisions of OPA apply to "each responsible party for a vessel or a facility from which oil is discharged" (33 U.S.C. § 2702). The responsible party is specifically tasked with further OPA obligations, including claim duties. Some have identified OPA's specific assignment of liability (often referred to as "channeling")[9] and other duties as a key component of the framework. The channeling mechanism may simplify the compensation process, because the responsible party assignment makes it unnecessary for agencies and courts to determine which party caused the spill.[10]

The term "responsible party" has a specific meaning for different sources of oil spills.[11] As defined by OPA (Section 1001), "responsible party" means the following:

a) Vessels. - In the case of a vessel, any person owning, operating, or demise chartering the vessel.
b) Onshore facilities. - In the case of an onshore facility (other than a pipeline), any person owning or operating the facility, except a Federal agency, State, municipality, commission, or political subdivision of a State, or any interstate body, that as the owner transfers possession and right to use the property to another person by lease, assignment, or permit.
c) Offshore facilities. - In the case of an offshore facility (other than a pipeline or a deepwater port licensed under the Deepwater Port Act of 1974 (33 U.S.C. § 1501 et seq.)), the lessee or permittee of the area in which the facility is located or the holder of a right of use and easement granted under applicable State law or the Outer Continental Shelf Lands Act (43 U.S.C. § 1301-1356) for the area in which the facility is located (if the holder is a different person than the lessee or permittee), except a Federal agency, State, municipality, commission, or political subdivision of a State, or any interstate body, that as owner transfers possession and right to use the property to another person by lease, assignment, or permit.
d) Deepwater ports. - In the case of a deepwater port licensed under the Deepwater Port Act of 1974 (33 U.S.C. § 1501- 1524), the licensee.
e) Pipelines. - In the case of a pipeline, any person owning or operating the pipeline.
f) Abandonment. - In the case of an abandoned vessel, onshore facility, deepwater port, pipeline, or offshore facility, the persons who would have been responsible parties immediately prior to the abandonment of the vessel or facility.

Liability

OPA unified the liability provisions of existing oil spill statutes,[12] creating a freestanding liability regime. Section 1002 states that responsible parties are liable for any discharge of oil (or threat of discharge) from a vessel or facility[13] to navigable waters, adjoining shorelines, or the exclusive economic zone of the United States (i.e., 200 nautical miles beyond the shore). Liability under OPA is strict,[14] meaning that impacted parties need not show or prove that the spiller acted negligently for liability to attach.[15]

Under OPA, a responsible party is liable for cleanup costs incurred, not only by a government entity, but also by a private party. But the cleanup activities must be consistent with the National Oil and Hazardous Substances Pollution Contingency Plan, generally referred to as the National Contingency Plan (NCP), the regulations governing oil and hazardous substance response operations.[16]

In addition, OPA broadened the scope of damages (i.e., costs) for which an oil spiller would be liable. (For a historical comparison of oil spill liability provisions, see Table A-1 in the Appendix to this report.) Damages include the following:

- **Natural resources:** "damages for injury to, destruction of, loss of, or loss of use of, natural resources, including the reasonable costs of assessing the damage, which shall be recoverable by a United States trustee, a State trustee, an Indian tribe trustee, or a foreign trustee."[17]
- **Real or personal property:** "damages for injury to, or economic losses resulting from destruction of, real or personal property, which shall be recoverable by a claimant who owns or leases that property."[18]
- **Subsistence use:** "damages for loss of subsistence use of natural resources, which shall be recoverable by any claimant who so uses natural resources which have been injured, destroyed, or lost, without regard to the ownership or management of the resources."[19]
- **Revenues:** "damages equal to the net loss of taxes, royalties, rents, fees, or net profit shares due to the injury, destruction, or loss of real property, personal property, or natural resources, which shall be recoverable by the Government of the United States, a State, or a political subdivision thereof."[20]
- **Profits and earning capacity:** "damages equal to the loss of profits or impairment of earning capacity due to the injury, destruction, or loss of real property, personal property, or natural resources, which shall be recoverable by any claimant."[21]
- **Public services:** "damages for net costs of providing increased or additional public services during or after removal activities, including protection from fire, safety, or health hazards, caused by a discharge of oil, which shall be recoverable by a State, or a political subdivision of a State."[22]

Liability and Compensation Issues Raised by the 2010 Gulf Oil Spill 7

OPA provided limited defenses from liability: Act of God, act of war, and act or omission of certain third parties.[23] These defenses are similar to those of the Comprehensive Environmental, Response, Compensation, and Liability Act (CERCLA), enacted in 1980 for releases of hazardous substances and pollutants or contaminants (but not oil).

Liability Limits

OPA provides liability limits (or caps) for those responsible for a spill. Liability limits are not unique to OPA and limits existed in several federal statutes preceding OPA. (For a historical comparison of liability limits see Table A-2 in the Appendix to this report.) However, the limits are not automatic, but conditional. First, the liability limits do not apply to situations involving acts of gross negligence or willful misconduct. Second, liability limits do not apply if the violation of a federal safety, construction, or operating requirement proximately caused the spill. Third, parties must report the incident and cooperate with response officials to maintain their liability caps.[24] According to the National Pollution Funds Center—an office of the U.S. Coast Guard that manages the Oil Spill Liability Trust Fund (discussed below)—liability limits are "not usually well defined until long after response," and litigation may be required to resolve the issue.[25]

The liability limits differ based on the source of the oil spill (*Table A-2*). The limits for most sources are simple dollar amounts.

- Vessel liability limits are generally based on the size of the vessel (measured in gross tonnage). For example, a tank vessel matching the size of the *Exxon Valdez* (95,000 gross tons) would have a cap of either $304 million (single-hull) or $190 million (double hull).
- Onshore facility (which includes pipelines) liability is limited to $350 million. Although OPA allows the President to decrease this limit through regulations, this authority has not been exercised.
- Deepwater port (e.g., Louisiana Offshore Oil Port, LOOP)[26] liability is limited to $350 million. OPA authorizes the Secretary of the department in which the Coast Guard is operating (i.e., Homeland Security)[27] to adjust this limit to not less than $50 million. This authority was exercised in 1995, setting the liability limit at $62 million,[28] and subsequently increased to $87 million in 2009.[29]
- Offshore facilities (like the BP oil well involved in the 2010 Gulf of Mexico spill) have unlimited liability for oil removal (cleanup) costs[30]

and a $75 million limit on other damages—natural resources and the five categories of economic damages (listed above).
- Mobile offshore drilling units (MODUs), like the *Deepwater Horizon*, are first treated as tank vessels for their liability cap. If removal and damage costs exceed this liability cap, a MODU is deemed to be an offshore facility for the excess amount.[31]

OPA requires the President to issue regulations to adjust the liability limits at least every three years to take into account changes in the consumer price index (CPI).[32] Despite this requirement, adjustments to liability limits were not made until Congress amended OPA in July 2006 (*Table A2*). As of the date of this report, onshore and offshore facility liability limits remain at the same level established in 1990. If the adjustments had been made, offshore facility liability limits for economic and natural resource damages would be approximately $125 million (plus unlimited removal costs).[33]

Financial Responsibility

To ensure that parties responsible for an oil spill can provide funding for oil spill response and compensation to affected parties, OPA requires that vessels and offshore facilities maintain evidence of financial responsibility (e.g., insurance or financial statements documenting significant revenue). OPA does not have an analogous requirement for onshore facilities.

The current levels of financial responsibility are related to the current liability limits for various sources (e.g., vessels, offshore facilities) of potential oil spills. The liability limits differ by potential source. In the case of vessels, whose liability limits are a single dollar amount encompassing both removal costs and other damages, the financial responsibility levels are directly tied to the corresponding liability caps. Current law requires responsible parties for vessels to demonstrate the "maximum amount of liability to which the responsible party could be subjected under [the liability limits in OPA Section 1004; 33 U.S.C. § 2704]."

Because the structure of offshore facility liability limit is different than vessels (liability for removal costs is unlimited), the corresponding financial responsibility limit provisions differ. Responsible parties for offshore facilities in federal waters must demonstrate $35 million financial responsibility, unless the President determines a greater amount (not to exceed $150 million) is justified (33 U.S.C. § 2716(c)). The federal regulations that implement this statutory provision (30 CFR Part 254) base the financial responsibility amount—between $35 million and $150 million—on a facility's worst-case discharge volume (as defined in 30 CFR Section 253.14). For example, a facility with a worst-case discharge volume over 105,000 barrels—the highest level of worst-case discharge listed in the regulations—must maintain $150 million in financial responsibility.

The Coast Guard's National Pollution Funds Center (NPFC) implements the financial responsibility provisions for vessels; the Bureau of Ocean Energy Management, Regulation, and Enforcement (formerly the Minerals Management Service, MMS) implements this requirement for offshore facilities.

The Oil Spill Liability Trust Fund

Prior to OPA's passage, a topic of debate concerned the mechanisms and hurdles of private parties recovering damages resulting after an oil spill.[34] To address this and other concerns,[35] Congress established the Oil Spill Liability Trust Fund (OSLTF). Although Congress created the OSLTF in 1986,[36] Congress did not authorize its use or provide its funding until after the 1989 *Exxon Valdez* oil spill. In 1990, OPA provided the statutory authorization necessary to put the fund in motion.[37]

In complementary legislation, Congress imposed a 5-cent-per-barrel tax on domestic and imported oil to support the fund.[38] Collection of this fee started January 1, 1990, and ceased on December 31, 1994, due to a sunset provision in the law. However, in April 2006, the tax resumed as required by the Energy Policy Act of 2005 (P.L. 109-58). In addition, the Emergency Economic Stabilization Act of 2008 (P.L. 110-343) increased the tax rate to 8 cents through 2016. In 2017, the rate is set to increase to 9 cents. The tax is scheduled to terminate at the end of 2017.[39]

The National Pollution Funds Center (NPFC), an office within the Coast Guard, manages the trust fund.[40] The trust fund plays a substantial role in the liability and compensation framework, as discussed below.

Compensation or Claims Process

OPA established a claims process for compensating parties affected by an oil spill. In general, before claims for removal costs and other costs/damages can be presented to the OSLTF, they must be presented first to a responsible party.[41] If the party to whom the claim is presented denies all liability, or if the claim is not settled by payment within 90 days after the claim was presented, the claimant may elect either to initiate an action in court against the responsible party or present the claim directly to the OSLTF.[42] If a responsible party denies a claim that is subsequently processed and awarded with monies from the OSLTF, the federal government may seek to recover these costs from the responsible party.[43]

Regulations implementing the OSLTF claims process are found in 33 C.F.R. Part 136. The NPFC's guidance document—*Claimant's Guide: A Compliance Guide for Submitting Claims Under the Oil Pollution Act of 1990* (November 2009)—provides assistance to those submitting claims.

OSLTF managers are limited in the amount of payments that may be awarded for each incident.[44] Under current law, the per-incident cap is $1 billion.[45] Because of this per-incident cap on the OSLTF, a scenario could arise in which the trust fund managers would be prohibited from compensating claimants who were initially denied by a responsible party. Such a scenario has not occurred in OPA's history.

Costs (including, for example, natural resource damages, economic losses, etc.)[46] beyond this per-incident limit could be addressed in several ways. One mechanism would be for parties to use state laws. OPA does not preempt states from imposing additional liability or requirements relating to oil spills, or establishing analogous state oil spill funds.[47] OPA legislative history[48] and statements from OPA drafters[49] indicate the intention that state laws and funds would supplement (if necessary) the federal liability framework under OPA.

Alternatively, existing federal authorities could be used to provide assistance in some circumstances. For example, an emergency declaration under the Stafford Act would appear to be a potential approach for the current situation, because it is intended to lessen the impact of an imminent disaster. Such a declaration in the context of a manmade disaster is unprecedented: during the *Exxon Valdez* spill, the President turned down the governor of Alaska's two requests for an emergency declaration.[50] CRS is not aware of similar requests made during the *Deepwater Horizon* incident.

ISSUES FOR POLICYMAKERS

The 2010 *Deepwater Horizon* oil spill generated considerable interest in the existing oil spill liability and compensation framework. The incident placed a spotlight on multiple elements of the framework, in particular the liability limits and the size and limitations of the OSLTF.

Liability and Compensation Issues Raised by the 2010 Gulf Oil Spill   11

The issues raised by the spill highlight a central policy debate: how should policymakers allocate the costs associated with a catastrophic oil spill? What share of costs should be borne by the responsible party (e.g., oil vessel owner/operators) compared to other groups, such as the oil industry (e.g., through the per-barrel tax), and/or the general treasury (assuming Congress would appropriate funds to compensate for unpaid costs/damages)?

Policymakers may want to consider the magnitude of the *Deepwater Horizon* incident and the liability and compensation issues raised under a scenario in which BP had refused to finance response activities or establish a claims process to comply with the relevant OPA provisions. BP has either directly funded oil spill response operations or reimbursed the federal government for actions taken by various agencies. According to BP, response costs have tallied over $10 billion.[51] BP has paid damage claims well above its liability limit of $75 million (assuming it would apply) and outside the scope of its liable damages (e.g., human health-related claims). BP and the Obama Administration jointly announced on June 16, 2010, the creation of the Gulf Coast Claims Facility (GCCF), an independent claims facility that BP will finance with incremental payments eventually totaling $20 billion.

Liability Limits

In the aftermath of the *Deepwater Horizon* spill, many Members of the 111th Congress expressed concern about the level of the liability limit for offshore facilities. Several Members offered proposals that would have either significantly increased the offshore facility liability limit or removed the limit entirely. These increases (e.g., to $10 billion) would have been well above the inflation adjustment increase required by OPA (if implemented would have raised the limit to $125 million). In July 2010, the House passed legislation that would have removed the liability limit for offshore facilities.[52] The Senate placed a comparable bill (S. 3663) on its Legislative Calendar in late July 2010, but did not vote on its passage. In the 112th Congress, several Members have offered proposals that would eliminate the liability limit for offshore facilities.[53]

Liability limits have been a part of the oil spill framework for decades (*Table A-2*). Eliminating the offshore facility liability limit altogether would constitute a substantial change in U.S. oil spill policy. In the current system, costs from a major spill are shared between the responsible party (an individual company) and the OSLTF (largely financed through a tax on oil).

Until the *Deepwater Horizon* incident, no individual spill threatened the framework.[54]

Unlimited liability for offshore facilities would shift the burden to the individual company. Although such a shift would likely reduce the risk of depleting the OSLTF during a catastrophic spill, the OSTLF would continue to provide multiple functions, including a role as a compensation backstop. Within a system of unlimited liability, the risk remains that the responsible party would fail to meets its compensation obligations (for whatever reason), and the OSLTF would continue to provide funds.

Indeed, raising or removing the liability limits would not guarantee that a company would be able to fund all response costs and compensate all affected parties. If Congress increases or eliminates the liability cap without making a corresponding change to the financial responsibility requirements, a responsible party could comply with its financial responsibility requirements and still go bankrupt before paying even a small fraction of the damage associated with a spill.[55]

Some proponents of increasing (or abolishing) the liability limit have argued that the current cap distorts economic decisions and provides incentives that may increase the likelihood of an oil spill.[56] Others point out that a liability cap represents a subsidy to the offshore oil industry—the lower the cap, the greater the subsidy.[57]

Others argue that a significant increase in the liability cap (or its removal) would be problematic from an insurance standpoint, depending, in part, on whether the financial responsibility requirements have corresponding increases.[58] Some contend that the insurance market does not have the capacity to meet a significant increase in the liability limit (e.g., to $10 billion).[59] These concerns are based on current market conditions, but the market may be able to adjust to different requirements. To what degree it can adjust is beyond the scope of this report, but some initial evidence suggests the market is adjusting. The January 2011 final report from the National Commission on the BP *Deepwater Horizon* Oil Spill and Offshore Drilling highlighted a September 2010 announcement from an insurance company (Munich Re) advertising coverage in the $10 billion to $20 billion range.[60]

Some have argued that higher liability limits would be a disadvantage to the companies that cannot self-insure (as BP has done), because insurance may become cost-prohibitive or even unobtainable (as argued above). Along this line of reasoning, increased liability provisions (and corresponding financial responsibility demonstrations) may preclude (relatively) smaller companies from offshore operations and increase the marketshare of the small group of major oil companies. Others counter that if companies cannot afford to bear the potential costs of their activities, they should not be in operation.[61]

OSLTF Limitations

Per-Incident Cap

Although reaching the OSLTF's per-incident ($1 billion) cap would be an unprecedented event in the fund's history, it is a conceivable occurrence with the *Deepwater Horizon* incident. As of November 7, 2010, the trust fund expenditures and obligations have exceeded $690 million.[62] As reported on the federal government's *Deepwater Horizon* response website,[63] BP has reimbursed the federal government through multiple payments, totaling over $600 million.[64] Although the reimbursement payments would be transferred into the OSLTF,[65] they have no effect on the trust fund expenditures and obligations and their relationship with the per-incident cap.[66]

Several proposals from the 111th Congress would have increased the per-incident cap to varying amounts. Often these proposals were coupled with provisions to bolster the revenues in the trust fund (discussed below). In the 112th Congress, several Members have offered proposals that would increase the per-incident cap.[67]

Level of Funding

The magnitude of the costs associated with *Deepwater Horizon* has spurred concern regarding the level of funding in the OSLTF. As *Figure 1* illustrates, the OSLTF unobligated balance was approximately of $1.7 billion at the end of FY2010.[68] The National Pollution Funds Center projects the fund will reach approximately $2.7 billion at the end of FY2014.[69] Although these projected levels are substantially higher than historic levels, they are unlikely to be sufficient to mitigate impacts—finance response activities and compensate injured parties—from a catastrophic spill akin to the *Deepwater Horizon* incident.

However, data from a 2007 GAO report suggest that the projected levels may be sufficient to address the more common mix of spills that have historically occurred. In 2007, GAO examined occurrences of vessel liability limits being exceeded and resulting trust fund vulnerability. GAO found that "major oil spills [defined by GAO as those with response costs and damage claims exceeding $1 million] that exceed the vessel's limit of liability are infrequent, but their impact on the Fund could be significant.... Of the 51 major oil spills that occurred since 1990 [which accounted for 2% of all oil spills since 1990], 10 spills resulted in limit of liability claims on the Fund."[70]

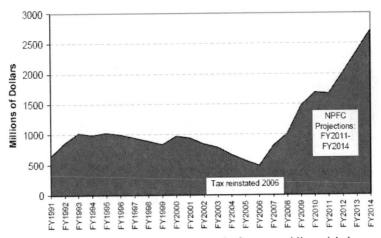

Source: Prepared by CRS; FY1991-FY2010 end-of-year unobligated balances from Office of Management and Budget, Budget of the United States Government, Appendix. FY2011-FY2014 projections from the National Pollution Funds Center, prepared January 18, 2011, and provided to CRS in personal communication February 2011.

Notes: Most of the trust fund's revenue has come from a per-barrel tax on domestic and imported oil. The Omnibus Budget Reconciliation Act of 1989 (P.L. 101-239) initiated the tax in January 1990 at 5 cents-per-barrel. Due to a sunset provision in the law, collection ceased December 31, 1994. In April 2006, the tax resumed as required by the Energy Policy Act of 2005 (P.L. 109-58). In 2008, pursuant to the Emergency Economic Stabilization Act of 2008 (P.L. 110-343), the tax rate increased to 8 cents through 2016. In 2017, the rate will increase to 9 cents, but is scheduled to terminate at the end of 2017.

The apparent plateau in the data between FY2010 and FY2011 is due to the transition between OMB observed values and NPFC projections. The NPFC projections include several assumptions: (1) Fines/penalties receipts based on recent annual averages; (2) claims based on recent averages; (3) 100% cost recovery for expenses associated with *Deepwater Horizon* incident.

Figure 1. Oil Spill Liability Trust Fund. Historical, Current, and Projected Balance.

In the 111th Congress, Members offered several proposals that would have increased the per-barrel tax that finances the OSLTF. Some of these proposals would have increased the rate substantially, and over time, provided substantial revenue to the OSLTF.[71] In the 112th Congress, Members have offered at least one bill that would alter the OSLTF tax rate.[72]

Potential CWA penalties may play a role in increasing the level of funding in the OSLTF.[73] Unless specifically addressed otherwise, the Miscellaneous Receipts Act (31 U.S.C. §3302(b)) provides that all court, or administratively imposed penalties are paid to the U.S. Treasury. The underlying statutory provisions of the OSLTF effectively override this general provision by transferring CWA Section 311 penalties (among others)[74] into the OSLTF.[75] CWA penalties from the *Deepwater Horizon* incident could be substantial. Sections 311(b)(7)(A) and 311(b)(7)(D)[76] provide for the maximum civil penalties under the CWA. The amount of revenue generated by applying these provisions depends on several key factors, including (1) the estimate of the amount of oil discharged;[77] (2) a determination of whether the oil spill was a result of "gross negligence or willful misconduct;"[78] (3) a decision of whether the oil that was directly captured—approximately 820,000 barrels—by BP via a pipeline-to-surface-vessel system is subtracted from the total estimate; and (4) the application of the CWA factors that must be considered by the "EPA Administrator, the Secretary [of Homeland Security], or the court, as the case may be," when determining penalty amounts.[79]

The recent National Commission report included a range of $4.5 billion to $21.5 billion for possible CWA penalty revenue.[80] Providing a more precise estimate is impossible, due to the four factors identified above. Moreover, several Members of Congress (with support from the Administration)[81] have offered legislation that would redirect any CWA *Deepwater Horizon* penalties to fund Gulf restoration efforts. The degree to which this goal is met will impact the level of funding in the OSLTF.

Claims Process

The OPA oil spill claims process has generated considerable interest in the aftermath of the *Deepwater Horizon* incident. BP is seeking to fulfill its OPA claims obligations through the Gulf Coast Claims Facility (GCCF, see text box below). Many have voiced concerns that claimants were/are not receiving adequate payments for their losses or receiving payments in a timely fashion.[82] The 111th Congress held at least four hearings on compensation and claims process issues.[83] Some interest has remained in the 112th Congress.[84]

Evaluating the validity of criticism expressed toward the GCCF presents challenges. A key reason for difficulty is the lack of information (i.e., transparency) available to assess the GCCF process. Indeed, some have pointed to the lack of transparency as a substantial concern.[85]

> **The Gulf Coast Claims Facility**
>
> In the early weeks of the spill response, the issue of BP's liability limit received considerable attention. On many occasions BP executives stated that the company would pay all "legitimate claims,"[86] but some Members of Congress were skeptical about BP's application of the term "legitimate." For example, would legitimate claims entail only those within BP's liability limit? A complicating factor in the debate was the uncertainty regarding whether BP would be subject to a liability limit. At least in part to address this uncertainty,[87] BP agreed (after discussions with the Obama Administration) to establish a $20 billion fund to compensate those affected by the spill.[88] The claims would be processed through an independent claims facility, administered by Kenneth Feinberg. This facility—the Gulf Coast Claims Facility (GCCF)—began accepting claims August 23, 2010. Prior to that time, BP processed and awarded claims. According to BP, the company awarded $399 million in claims from May 3 until the August 23 transition to the GCCF.[89]

To address transparency concerns, the GCCF offered its final claim payment methodology for public comment February 2, 2011. The final claim methodology includes several assumptions about the expected recovery of the Gulf ecosystem and economy. For example, the GCCF states that "it is anticipated that, for all businesses other than oyster harvesting, recovery will continue in 2011 with full recovery expected in 2012."[90] Final claim payments are based on this future losses assumption, which the GCCF states will be reassessed every four months and thus may decrease/increase over time. It is unclear how the future losses assumption was derived.[91] The final methodology includes an expert opinion (as an appendix) which provides some basis for the assumption.[92] Regardless of one's perspective concerning this assumption, payment of final claims (that award claimants with expected losses in upcoming years) necessitates some projection of the timing of ecosystem recovery and associated economic losses.

The GCCF does provide some statistical data on its website,[93] allowing for a limited evaluation.

For example, GCCF data indicate that (as of February 10, 2011) approximately 38% of submitted claims have been paid, with percentages varying by state (*Table 1*).

Table 1. Deepwater Horizon Oil Spill Claims Data by State
As of February 10, 2011—States Listed In Order of Claims Submitted

	Louisiana	Florida	Alabama	Mississippi	Texas	Other	Total
Number of claims submitted	270,000	210,938	94,180	69,506	12,015	23,561	680,200
Number of claims paid	89,767	94,035	37,521	22,014	3,237	9,040	255,614
Percent of submitted claims receiving some level of payment	33%	45%	40%	32%	27%	38%	38%
Amount paid	$1,099,355,733	$1,205,226,776	$558,111,338	$281,526,529	$96,026,300	$134,083,900	$3,374,330,576

Source: Prepared by CRS with data from the GCCF website, at http://www.gulfcoastclaimsfacility.com/.

Notes: Above data reflect claims by state of residence (as opposed to state of loss). Above data only include claims paid through the GCCF, which began processing claims August 23, 2010. Prior to that date, BP paid approximately $400 million in claims.

Some have argued that the percent of claims receiving some payment (38%) may be overstating the effectiveness of the GCCF to pay out claims.[94] The basis for this argument is that a substantial portion of the paid claims (86,835 or 34%) are so-called "quick payments." These are expedited final claims (i.e., additional documentation not required) for individuals ($5,000) and businesses ($25,000) who received emergency advance payments (or would qualify for interim payments). If these expedited claims are not included in the data in *Table 1*, the percent of total paid claims decreases to 29%.

A key piece of information that emerged in a January 2011 Senate hearing raised the issue of whether BP (until August 23, 2010) and the GCCF (on and after August 23, 2010) have processed claims in accordance with procedures of the NPFC. During the hearing, the Director of the NPFC (Craig Bennett) stated that the NPFC has received approximately 500 claims related to the Gulf spill.[95] The Director stated that the NPFC has adjudicated 200 of these claims, all of which were denied. In the hearing this fact was presented as evidence that the GCCF was processing claims in accordance with NPFC protocols. However, the status of the claims that have not been adjudicated was not discussed. Without this information, an assessment of the BP/GCCF claims process and its consistency with the NPFC would be premature.

Although OPA does provide authority to the Comptroller General to audit claim activities of the NPFC,[96] OPA does not provide a federal agency with the authority to oversee or audit the claims activities of the responsible party. At least one proposal from the 111th Congress included provisions that would have addressed this issue.[97]

A related concern shared by some Members of Congress is the time period embedded in the OPA claims process. Before seeking compensation from the OSLTF, claimants must first submit their claim to the responsible party. If the claim is not settled within 90 days, the claimant may either pursue the claim through the OSLTF or seek recourse through litigation. Many argued that the 90- day clock was too long for impacted parties to wait, and several proposals from the 111th Congress would have altered the waiting period (many shortening the clock to 45 days).[98] At least one proposal in the 112th Congress would amend this provision.[99]

During the GCCF claims process, several stakeholders have criticized the "proximate cause" language of various GCCF protocols, arguing that such provisions are not consistent with OPA and its claims process.[100] The issue has evolved, with the GCCF Administrator addressing the concerns to some degree. The GCCF's August 2010 Emergency Advance Payment Protocol text read:

> The GCCF will only pay for harm or damage that is proximately caused by the Spill. The GCCF's causation determinations of OPA claims will be guided by OPA and federal law interpreting OPA and the proximate cause doctrine. Determinations of non-OPA claims will be guided by applicable law. The GCCF will take into account, among other things, geographic proximity, nature of industry, and dependence upon injured natural resources.[101]

In October 2010, Administrator Feinberg announced the GCCF was removing the geographic proximity text (referred to as the "geographic test") for claim eligibility that is part of the above protocol. The interim and final claim protocol now reads:

> The GCCF will only pay for harm or damage that is proximately caused by the Spill. The GCCF's causation determinations of OPA claims will be guided by OPA and federal law interpreting OPA. Determinations of physical injury and death claims will be guided by applicable law.

The relevant OPA provision does not specifically address geographic proximity, but states that "each responsible party for a vessel or a facility from which oil is discharged, or which poses the threat of a discharge of oil is liable for the removal costs and damages ... that *result from* such an incident."[102] Moreover, neither the implementing regulations (33 C.F.R. Part 136) nor OSLTF claim guidance documents indicates that proximate cause is a consideration. Of course, as a practical matter the OSLTF must apply *some* cut-off akin to proximate cause so that injuries far down the causal chain of events following a spill are not compensated.

On this matter, the Presidential Deepwater Commission concluded:

> There is no easy legal answer to the question of how closely linked those lost profits or earnings must be to the spill before they should be deemed compensable. The search for such a rational endpoint for liability has already stymied the Gulf Coast Claims Facility in its processing of claims.[103]

POTENTIAL OPTIONS FOR POLICYMAKERS

The current combination of liability limits and $1 billion per-incident cap is not sufficient to withstand a spill with damages/costs that exceed a responsible party's liability limit (assuming it would apply) by $1 billion. Even if the per-incident cap were increased, the current (and projected) level of funds in the OSLTF may not be sufficient to address costs from a catastrophic spill.

The options available to address these issues depend upon on the overall objective of Congress. One objective—which has been expressed by many in and outside Congress—is to provide full restoration and timely compensation for the impacts from a catastrophic spill, without directly burdening the general taxpayers. In the context of this objective, "timely" compensation means that an injured party would have access to compensation without going through a court system, which would likely require more of a claimant's time and resources. With that objective in mind, several options are discussed below.

Potential options for Congress include (but are not limited to) the following, many of which were proposed in various legislation in the 111th Congress and some of which are included in proposals in the 112th Congress:[104]

- Increase the liability limits, so that the responsible party would be required to pay a greater portion of the total spill cost before accessing trust fund dollars. Congress may consider different limits for different offshore activities. Precedent exists in OPA for setting different liability limits to account for different oil spill risks: The liability limit for single-hulled tank vessels is approximately 50% higher than for double-hulled vessels. In the outer continental shelf (OCS) oil exploration and development sector, policymakers may consider a wide array of factors that could influence (1) the risk of an oil spill occurring and (2) the risk that the oil spill could not be contained before impacting sensitive ecosystems and/or affecting large populations. Policymakers could then structure the liability limit framework based on certain behavior, the use of specific technologies, and/or the location of the activity. However, CRS is not aware of a comprehensive risk assessment of individual factors (or their combinations) regarding OCS drilling activities. A rigorous analysis of possible risk factors could be instructive to policymakers.[105]
- Increase the required financial responsibility coverage, either matching the increased liability limit (as OPA requires for vessels) or setting the coverage at a level based on other factors, such as capacity in the insurance market. Congress could increase coverage amounts

through a staggered approach, to allow more time for the market to adjust.
- Remove or raise the per-incident cap on the trust fund. If removed entirely, the fund could be at risk of depletion with one incident.
- Increase the per-barrel oil tax to more quickly raise the fund's balance.
- Authorize "repayable advances" to be made (via the appropriations process) to the trust fund, so that the fund would have the resources to carry out its functions (cleanup efforts, claim awards). Up until 1995, the fund had this authority, in order to ensure it could respond to a major spill before the fund had an opportunity to grow (via the per-barrel tax).
- Require industry to establish a pool of funds (of significant magnitude) that would be available to finance response actions, injured party compensation, or both. Such a fund could either replace or complement the existing OPA system of individual liability and support from the OSLTF. This would be analogous to the framework for the nuclear power industry created by the Price-Anderson Act (primarily Section 170 of the Atomic Energy Act of 1954, 42 U.S.C. § 2210).[106]

Another objective might be to maintain the existing system, which may be sufficient to address all but the most extreme scenarios. Catastrophic spills in U.S. waters have historically been rare. Some may argue that establishing a system that can withstand a catastrophic event would impose costs and yield consequences that would not justify the (expected) ability to address a catastrophic event.

Interest in issues raised by the 2010 Gulf oil spill has waned in recent months. However, Members in the 112th Congress have introduced multiple proposals, many of which would address liability and compensation framework issues.[107] Legislative activity in the 112th Congress may be influenced by several factors, including (but not limited to) assessments of conditions in the Gulf region, reports from other independent inquiries,[108] further information regarding the claims process, and results from the natural resource damage assessment process. Moreover, it may be worth noting that passage of the Oil Pollution Act of 1990 occurred 18 months after the *Exxon Valdez* spill.

APPENDIX. LIABILITY TABLES — HISTORICAL PERSPECTIVES

Table A-1. Evolution of the Scope of Oil Spill Liability

Pre-OPA Clean Water Act[a]	Pre-OPA Outer Continental Shelf Lands Act Amendments[b]	Pre-OPA Deepwater Port Act[c]	Pre-OPA Trans-Alaska Pipeline Authorization Act[d]	OPA 1990
Liability applied to vessels and facilities that discharge oil into or on U.S. navigable waters, adjoining shorelines, and other specified areas.	Liability applied to offshore facilities and vessels transporting oil from offshore facilities.	Liability applied to deepwater ports and vessels operating in their vicinity. Applicable parties liable for cleanup costs and for damages that result from a discharge of oil.	Liability applied to the holder of a pipeline right-of-way and vessels transporting oil from the trans-Alaska pipeline. The vessel was liable for "all damages, including clean-up costs." The right-of-way holder was "liable to all damaged parties, public or private."	Liability applies to vessels and facilities that discharge oil into or upon navigable waters or adjoining shorelines or the exclusive economic zone.
Pre-OPA Clean Water Act[a]	Pre-OPA Outer Continental Shelf Lands Act Amendments[b]	Pre-OPA Deepwater Port Act[c]	Pre-OPA Trans-Alaska Pipeline Authorization Act[d]	OPA 1990

Pre-OPA Clean Water Act[a]	Pre-OPA Outer Continental Shelf Lands Act Amendments[b]	Pre-OPA Deepwater Port Act[c]	Pre-OPA Trans-Alaska Pipeline Authorization Act[d]	OPA 1990
Applicable parties liable for removal costs and natural resource damages.[e]	Applicable parties liable for removal costs, natural resource damages, and economic damages, including (1) real or personal property; (2) loss of use of real or personal property; (3) subsistence use; (4) profits and earning capacity; and (5) tax revenue.	"Damages" is defined to as "all damages (except cleanup costs) suffered by any person, or involving real or personal property, the natural resources of the marine environment, or the coastal environment of any nation, including damages claimed without regard to ownership of any affected lands, structures, fish, wildlife, or biotic or natural resources."		Applicable parties liable for removal costs, natural resource damages (and their assessments), and economic damages, including (1) real or personal property; (2) subsistence use; (3) public revenues; (4) profits and earning capacity; and (5) public services.

Source: Prepared by CRS.

Notes:

[a] The Clean Water Act contains liability limits for vessels and facilities in §311(f) (33 U.S.C. §1321(f)) that were in place before OPA's enactment in 1990. These provisions remain. However, OPA §2002 states that the CWA provisions do not apply to incidents that fall under OPA's liability.

[b] Outer Continental Shelf Lands Act Amendments of 1978 (OCSLAA, 43 U.S.C. § 1801 et seq). OPA Section 2003 repealed the liability provisions in the OCSLAA (43 U.S.C. §§1811-1824).

[c] Deepwater Port Act of 1974 (DWPA, 33 U.S.C. § 1501 et seq). OPA Section 2003 repealed the liability provisions in the Deepwater Port Act (33 U.S.C. §1517).

[d] Trans-Alaska Pipeline Authorization Act of 1973 (TAPAA, 43 U.S.C. § 1651). OPA §§ 8101-8102 amended provisions of TAPAA, repealing 43 U.S.C. §1653(c), which applied to vessels transporting oil from the trans-Alaska pipeline.

[e] 33 U.S.C. §1321(f)(4) states that removal costs include the costs of restoration or replacement of natural resources damaged by an oil spill.

Table A-2. Evolution of Oil Spill Liability Limits

	Pre-OPA CWA[a]	Other Pre-OPA Statutes[b]	OPA 1990	Current Limit
Tank Vessels/Barges				
-Single-hull	The greater of: $125/gross ton or $125,000 for inland oil barges; $150/gross ton or $250,000 for other tank vessels/barges	OCSLAA: the greater of $300/gross ton or $250,000[c] DWPA: the lesser of $150/gross ton or $20 million TAPAA: $14 million	The greater of $1,200/gross ton or (1) $10 million if vessel is more than or equal to 3,000 gross tons; (2) $2 million if vessel is less than 3,000 gross tons	The greater of $3,200/gross ton or (1) ~$23.5 million if vessel is more than or equal to 3,000 gross tons (2) ~$6.4 million if vessel is less than 3,000 gross tons.[d]
-Double-hull	Same as single-hull	Same as single-hull	Same as single-hull	The greater of $2,000/gross ton or (1) ~$17.1 million if vessel is more than or equal to 3,000 gross tons (2) ~$4.3 million if vessel is less than 3,000 gross tons.[e]
Non-Tank Vessels	$150/gross ton	NA	The greater of $600/gross ton or $500,000	The greater of $1,000/gross ton or $854,000.[f]
Offshore Facilities	$50 million	OCSLAA: $35 million for natural resource damages and covered economic damages; removal costs not limited.	$75 million for natural resource damages and covered economic damages; removal costs not limited	Same as OPA 1990[g]

	Pre-OPA CWA[a]	Other Pre-OPA Statutes[b]	OPA 1990	Current Limit
Onshore Facilities (includes pipelines)	$50 million	TAPAA: Trans-Alaska pipeline right-of-way holder liable for "total removal of the pollutant." Liability limited to $50 million for other damages.	$350 million for removal costs, natural resource damages, and covered economic damages; allows President to decrease limit through regulations, but this authority has not been exercised.	Same as OPA 1990h
Deepwater Ports Louisiana Offshore Oil Port (LOOP)i	$50 millionj	DWPA: $50 million	$350 million for removal costs, natural resource damages, and covered economic damages, but allowed the Secretary (of the department in which the Coast Guard operates)k to adjust this limit to not less than $50 million. In 1995, the Department of Transportation set the liability limit for the LOOP at $62 million.l	$87.6 million for the LOOP.m

Source: Prepared by CRS.

Notes:

[a] The Clean Water Act contains liability limits for vessels and facilities in §311(f) (33 U.S.C. §1321(f)). These were in place before OPA's enactment in 1990 and were not repealed by OPA. However, OPA §2002 states that the CWA provisions do not apply to incidents that fall under OPA's liability.

[b] In addition to the CWA, three other statutes had liability provisions that applied to some oil spills before OPA's passage: the Trans-Alaska Pipeline Authorization Act of 1973 (TAPAA, 43 U.S.C. § 1651), the Deepwater Port Act of 1974 (DPA, 33 U.S.C. § 1501 et seq), and the Outer Continental Shelf Lands Act Amendments of 1978 (OCSLAA, 43 U.S.C. § 1801 et seq). These statutes were limited to covering special spills related to specific oil spill events. OPA Section 2003 repealed the liability provisions in the Deepwater Port Act (33 U.S.C. §1517) and the OCSLAA (43 U.S.C. §§1811-1824). OPA §§ 8101-8102 amended provisions of TAPAA, repealing 43 U.S.C. §1653(c), which applied to vessels transporting oil from the trans-Alaska pipeline.

[c] OCSLAA §1814 (repealed by OPA §2003).
[d] The Coast Guard and Maritime Transportation Act of 2006 (P.L. 109-241) increased limits to $1,900/gross ton for double-hulled vessels and $3,000/gross ton for single-hulled vessels. The Coast Guard made further adjustments (pursuant to the consumer price index provision in OPA §1004(d)(4)) in 2009: The limits are codified in 33 CFR §138.230.
[e] 33 CFR §138.230.
[f] 33 CFR §138.230.
[g] Although OPA §1004(d)(4) requires the President to issue regulations to increase the liability limit (per the consumer price index), the offshore facility limit has remained constant.
[h] Although OPA §1004(d)(4) requires the President to issue regulations to increase the liability limit (per the consumer price index), the onshore facility limit has remained constant.
[i] There is only one deepwater port for oil in U.S. coastal waters: the Louisiana Offshore Oil Port (LOOP). According to the U.S. Department of Transportation's Maritime Administration, three other deepwater ports are in operation that accept liquefied natural gas. See httpi/www.marad.dot.gov.
[j] Deepwater ports are not specifically identified in CWA §311(f), but would likely meet the definition of offshore facility.
[k] The Homeland Security Act of 2002 (P.L. 107-296) transferred the Coast Guard to the Department of Homeland Security. The Coast Guard was formerly located within the Department of Transportation.
[l] 60 *Federal Register* 39849, August 4, 1995.
[m] 33 CFR §138.230(b).

End Notes

[1] A portion (17%) of this oil did not enter the Gulf environment, but was directly recovered by BP. See, Federal Interagency Solutions Group, Oil Budget Calculator Science and Engineering Team, *Oil Budget Calculator: Deepwater Horizon-Technical Documentation*, November 2010.

[2] Transocean owned the *Deepwater Horizon* drilling rig. Three companies own the Macondo well: BP has a 65% share, Anadarko Petroleum Corporation has a 25% share, and MOEX Offshore has a 10 % share (National Commission on the BP Deepwater Horizon Oil Spill and Offshore Drilling, Deep Water: The Gulf Oil Disaster and the Future of Offshore Drilling, Report to the President (hereinafter "Commission Report") January 2011).

[3] In addition, this figure does not include punitive damages. Punitive damage claims were litigated for more than 12 years, eventually reaching the U.S. Supreme Court in 2008 (*Exxon Shipping v. Baker*, 554 U.S. 471 (2008)). Plaintiffs were awarded approximately $500 million in punitive damages—substantially less than was originally awarded ($5 billion) by a U.S. district court in 1994. An additional $500 million in post-judgment interest on those damages was subsequently awarded.

[4] As stated by BP: "The total amounts that will ultimately be paid by BP in relation to all obligations relating to the incident are subject to significant uncertainty and the ultimate exposure and cost to BP will be dependent on many factors. Furthermore, the amount of claims that become payable by BP, the amount of fines ultimately levied on BP (including any determination of BP's negligence), the outcome of litigation, and any costs arising from any longer-term environmental consequences of the oil spill, will also impact upon the ultimate cost for BP." BP, Fourth quarter and full year 2010 financial statement, February 1, 2011, at http://www.bp.com/.

[5] In the 111th Congress, the House of Representatives conducted at least 33 hearings in 10 committees; the Senate conducted at least 30 hearings in 8 committees. Members introduced more than 150 legislative proposals that would address various topics related to the oil spill.

[6] See CRS Report R41684, *Oil Spill Legislation in the 112th Congress*, by Jonathan L. Ramseur.

[7] P.L. 101-380, primarily codified at 33 U.S.C. § 2701 *et seq*. OPA amended other sections of the U.S. Code, including the Clean Water Act (e.g., 33 U.S.C. § 1321) and portions of the tax code (26 U.S.C. § 4611 and § 9509).

[8] See also CRS Report R41266, *Oil Pollution Act of 1990 (OPA): Liability of Responsible Parties*, by Robert Meltz; and CRS Report RL33705, *Oil Spills in U.S. Coastal Waters: Background and Governance*, by Jonathan L. Ramseur.

[9] See e.g., Nathan Richardson, *Deepwater Horizon and the Patchwork of Oil Spill Liability Law*, Resources for the Future discussion paper, June 2010.

[10] See e.g., Mark A. Cohen et al., *Deepwater Drilling: Law, Policy, and Economics of Firm Organization and Safety*, Resources for the Future discussion paper, January 2011.

[11] See 33 U.S.C. § 2701(32).

[12] The Federal Water Pollution Control Act of 1972, as amended, generally referred to as the Clean Water Act (CWA, 33 U.S.C. § 1251 et seq); the Trans-Alaska Pipeline Authorization Act of 1973 (TAPAA, 43 U.S.C. § 1651), the Deepwater Port Act of 1974 (DPA, 33 U.S.C. § 1501 et seq), and the Outer Continental Shelf Lands Act Amendments of 1978 (OCSLAA, 43 U.S.C. § 1801 et seq).

[13] The definition of "facility" is broadly worded and includes pipelines and motor vehicles. OPA § 1001(9).

[14] Under OPA, the terms "liable" and "liability" are "construed to be the standard of liability which obtains under section 311 of the [Clean Water Act]." Courts have interpreted section 311 of the Clean Water Act as imposing strict liability on parties responsible for the discharge of oil or hazardous substances into the waters of the United States. See United States v. New York, 481 F. Supp. 4 (S. D.N.Y. 1979). See CRS Report R41266, *Oil Pollution Act of 1990 (OPA): Liability of Responsible Parties*, by Robert Meltz.

[15] See, e.g., Nathan Richardson, *Deepwater Horizon and the Patchwork of Oil Spill Liability Law*, Resources for the Future discussion paper, June 2010.

[16] OPA §1002(b)(1)(B). The NCP is codified in 40 CFR Part 300. For further information on the NCP see CRS Report RL33705, *Oil Spills in U.S. Coastal Waters: Background and Governance*, by Jonathan L. Ramseur.

[17] OPA § 1002(b)(2)(A).

[18] OPA § 1002(b)(2)(B).

[19] OPA § 1002(b)(2)(C).

[20] OPA § 1002(b)(2)(D).

[21] OPA § 1002(b)(2)(E).

[22] OPA § 1002(b)(2)(F).

[23] OPA § 1003.

[24] OPA § 1004(c).

[25] National Pollution Funds Center, *FOSC Funding Information for Oil Spills and Hazardous Materials Releases*, April 2003, p. 4.

[26] The Louisiana Offshore Oil Port (LOOP) is the only offshore deepwater port for oil in U.S. coastal waters. According to the U.S. Department of Transportation's Maritime Administration, three other deepwater ports are in operation that accept liquefied natural gas. See http://www.marad.dot.gov.

[27] The Homeland Security Act of 2002 (P.L. 107-296) transferred the Coast Guard to the Department of Homeland Security. The Coast Guard was formerly within the Department of Transportation.

[28] 60 *Federal Register* 39849, August 4, 1995.

[29] 74 *Federal Register* 31368, July 1, 2009. Codified in 33 CFR § 138.230(b).

[30] OPA § 1002 defines removal costs as "the costs of removal that are incurred after a discharge of oil has occurred or, in any case in which there is a substantial threat of a discharge of oil, the costs to prevent, minimize, or mitigate oil pollution from such an incident." Relatedly, OPA § 1002 defines remove or removal as "containment and removal of oil or a hazardous substance from water and shorelines or the taking of other actions as may be necessary to minimize or mitigate damage to the public health or welfare, including, but not limited to, fish, shellfish, wildlife, and public and private property, shorelines, and beaches."

[31] 33 USC § 2704(b). For further interpretation see National Pollution Funds Center, "Oil Pollution Act Liabilities for Oil Removal Costs and Damages as They May Apply to the Deepwater Horizon Incident" (undated).

[32] With Executive Order 12777 (October 18, 1991), President George H.W. Bush delegated this responsibility to several federal agencies. Executive Order 13286 (signed by President George W. Bush on March 5, 2003) reorganized duties in response to the creation of the Department of Homeland Security. The Coast Guard covers vessels, deepwater ports (including associated pipelines), and transportation-related onshore facilities; the Department of Transportation (DOT) covers onshore pipelines, motor carriers, and railways; the Environmental Protection Agency (EPA) covers nontransportation-related

onshore facilities; the Department of the Interior (DOI) covers offshore facilities and associated pipelines, other than deepwater ports.

[33] CRS prepared this estimate using Consumer Price Index figures from the U.S Department of Labor Bureau of Labor Statistics, at ftp://ftp.bls.gov/pub/special.requests/cpi/cpiai.txt.

[34] See, e.g., U.S. Congress, House Committee on Merchant Marine and Fisheries, Report accompanying H.R. 1465, Oil Pollution Prevention, Removal, Liability, and Compensation Act of 1989, 1989, H.Rept. 101-242, Part 2, 101st Cong., 1st sess., p. 35.

[35] Another key function of the OSLTF is to provide a source of immediately accessible funds to support federal agencies, such as the Coast Guard or EPA, conducting oil spill response activities. Access to such funds is especially important if the responsible party is unknown or is unwilling or unable to pay for response activities.

[36] Omnibus Budget Reconciliation Act of 1986 (P.L. 99-509).

[37] Pursuant to OPA authorization, Congress transferred monies from other federal liability funds into the OSLTF, including the CWA Section 311(k) revolving fund; the Deepwater Port Liability Fund; the Trans-Alaska Pipeline Liability Fund; and the Offshore Oil Pollution Compensation Fund. According to the trust fund managers (the National Pollution Funds Center), no additional funds remain to be transferred to the OSLTF. National Pollution Funds Center, *Oil Spill Liability Trust Fund (OSLTF) Annual Report FY 2004–FY 2008*.

[38] Omnibus Budget Reconciliation Act of 1989 (P.L. 101-239). Other revenue sources for the fund include interest on the fund, cost recovery from the parties responsible for the spills, and any fines or civil penalties collected.

[39] Section 405 of P.L. 110-343.

[40] Pursuant to Executive Order 12777, 56 *Federal Register* 54,757 (1991).

[41] 33 U.S.C. § 2713(a). Under OPA, the term "claim" means "a request, made in writing for a sum certain, for compensation for damages or removal costs resulting from an [oil spill] incident." 33 U.S.C. § 2701(3).

[42] 33 U.S.C. § 2713(c). Claims for removal costs must be presented to the Fund within six years after the date of completion of all removal activities related to the oil spill incident. 33 U.S.C. § 2712(h)(1). Damage claims must be "presented within 3 years after the date on which the injury and its connection with the discharge in question were reasonably discoverable with the exercise of due care." 33 U.S.C. § 2712(h)(2).

[43] The federal government may also seek cost recovery from the responsible party's guarantor, "or any other person who is liable, pursuant to any law, to the compensated claimant or to the Fund, for the cost or damages for which the compensation was paid." 33 U.S.C. § 2715(c).

[44] "Incident" means any occurrence or series of occurrences having the same origin, involving one or more vessels, facilities, or any combination thereof, resulting in the discharge or substantial threat of discharge of oil. 33 U.S.C. § 2701(14).

[45] 26 U.S.C. § 9509(c).

[46] Although offshore facilities are liable for all removal costs, liability for removal costs for other responsible party categories (e.g., tank vessels, onshore facilities) is limited. Thus, a significant oil spill from a tank vessel could potentially encounter the per-incident trust fund cap, based solely on its response costs.

[47] 33 U.S.C. § 2718.

[48] U.S. Congress, House Committee on Merchant Marine and Fisheries, Report accompanying H.R. 1465, Oil Pollution Prevention, Removal, Liability, and Compensation Act of 1989, 1989, H.Rept. 101-242, Part 2, 101st Cong., 1st sess., p. 36.

[49] See George Mitchell, "Preservation of State and Federal Authority under the Oil Pollution Act of 1990," *Environmental Law*, Vol. 21, no. 2 (1991).

[50] The rationale for the turndowns was that a declaration by the President would hinder the government's litigation against Exxon that promised substantial compensation for the incident. See CRS Report R41234, *Potential Stafford Act Declarations for the Gulf Coast Oil Spill: Issues for Congress*, by Francis X. McCarthy.

[51] For details, see BP, Fourth quarter and full year 2010 financial statement, February 1, 2011, at http://www.bp.com/.

[52] H.R. 3534, the Consolidated Land, Energy, and Aquatic Resources Act (CLEAR Act), passed July 30, 2010. For more information, see CRS Report R41453, *Oil Spill Legislation in the 111th Congress*, by Jonathan L. Ramseur.

[53] See CRS Report R41684, *Oil Spill Legislation in the 112th Congress*, by Jonathan L. Ramseur.

[54] A 2007 GAO report examined occurrences of vessel liability limits being exceeded and resulting trust fund vulnerability. GAO found that "major oil spills [defined by GAO as those with response costs and damage claims exceeding $1 million] that exceed the vessel's limit of liability are infrequent, but their impact on the Fund could be significant.... Of the 51 major oil spills that occurred since 1990 [which accounted for 2% of all oil spills since 1990], 10 spills resulted in limit of liability claims on the Fund." GAO, *Maritime Transportation: Major Oil Spills Occur Infrequently, but Risks to the Federal Oil Spill Fund Remain*, Sept. 7, 2007.

[55] National Commission on the BP Deepwater Horizon Oil Spill and Offshore Drilling, Liability and Compensation Requirements under the Oil Pollution Act, Staff Working Paper No. 10, March 2011.

[56] Michael Greenstone (Massachusetts Institute of Technology), Testimony before the House Committee on Transportation and Infrastructure, June 9, 2010.

[57] Thomas W. Merrill (Columbia Law School), "Insurance and Safety Incentives," Working Paper for the National Commission on the BP Deepwater Horizon Oil Spill and Offshore Drilling, January 2011.

[58] For a more thorough discussion of these issues, see CRS Report R41320, *Deepwater Horizon Oil Spill Disaster: Risk, Recovery, and Insurance Implications*, by Rawle O. King.

[59] Robert P. Hartwig (Insurance Information Institute), Testimony before the House Committee on Transportation and Infrastructure, June 9, 2010; Ron Baron (Willis, Global Energy Practice), Testimony before the Senate Committee on Environment and Public Works, June 9, 2010.

[60] Press Release from Munich Re, September 12, 2010, at http://www.munichre.com/en/media press_releases/2010/2010_09_12jress_release.aspx.

[61] Se e.g., Thomas W. Merrill (Columbia Law School), "Insurance and Safety Incentives," Working Paper for the National Commission on the BP Deepwater Horizon Oil Spill and Offshore Drilling, January 2011.

[62] Personal communication with the U.S. Coast Guard, November 22, 2010.

[63] See http://www.restorethegulf.gov.

[64] According to the federal government's response website (as of January 11, 2011), BP has provided 8 reimbursement payments totaling $606 million. A further bill for over $25 million is pending. See http://www.restorethegulf.gov/ release/2011/01/11/oil-spill-cost-and-reimbursement-fact-sheet.

[65] Pursuant to 26 U.S.C. § 9509(b)(3), referencing OPA § 1015 (33 U.S.C. § 2715), which concerns cost recovery actions.

[66] 33 U.S.C. § 9509(c)(2). See also GAO, *Deepwater Horizon Oil Spill: Preliminary Assessment of Federal Financial Risks and Cost Reimbursement and Notification Policies and Procedures: Briefing for Congressional Requesters*, November 9, 2010.

[67] See CRS Report R41684, *Oil Spill Legislation in the 112th Congress*, by Jonathan L. Ramseur.

[68] Office of Management and Budget, *Budget of the U.S. Government for Fiscal Year 2012*, Appendix, pp. 529-530.

[69] Personal communication with the NPFC, February 2011.

[70] GAO, *Maritime Transportation: Major Oil Spills Occur Infrequently, but Risks to the Federal Oil Spill Fund Remain*, Sept. 7, 2007.

[71] For example, S. 3793 (Baucus) would have increased the tax from 8 to 78 cents, raising $31.4 billion over 10 years (Summary of the Baucus Job Creation and Tax Cuts Act, September 16, 2010).

[72] See CRS Report R41684, *Oil Spill Legislation in the 112th Congress*, by Jonathan L. Ramseur.

[73] See CRS Report R41370, *Federal Civil and Criminal Penalties Possibly Applicable to Parties Responsible for the Gulf of Mexico Oil Spill*, by Robert Meltz.

[74] 26 U.S.C. § 9509(b)(8) states "any penalty paid pursuant to section 311 of the Federal Water Pollution Control Act, section 309(c) of such Act (as a result of violations of such section 311), the Deepwater Port Act of 1974, or section 207 of the Trans-Alaska Pipeline Authorization Act."

[75] The relationship between trust funds, such as the OSLTF, and the general treasury is complex and beyond the scope of this report. For more information, see GAO, *Federal Trust and Other Earmarked Funds: Answers to Frequently Asked Questions*, January 2001.

[76] 33 U.S.C. § 1321(b)(7); see also 40 C.F.R. § 19.4, which increased various penalty amounts to account for inflation.

[77] See CRS Report R41531, *Deepwater Horizon Oil Spill: The Fate of the Oil*, by Jonathan L. Ramseur.

[78] Such a determination would increase the maximum per-barrel penalty from $1,100/barrel to $4,300/barrel.

[79] As listed in CWA § 311(b)(8), these include "the seriousness of the violation or violations, the economic benefit to the violator, if any, resulting from the violation, the degree of culpability involved, any other penalty for the same incident, any history of prior violations, the nature, extent, and degree of success of any efforts of the violator to minimize or mitigate the effects of the discharge, the economic impact of the penalty on the violator, and any other matters as justice may require."

[80] See Commission Report, p. 211. The low end of this range is achieved by multiplying 4.1 million barrels (amount of discharge after removing the 17% directly captured by BP) by $1,100/ barrel. The upper end of range is achieved by multiplying 4.9 million barrels (total discharge amount) by the maximum penalty of $4,300/barrel, which presumes a determination of either gross negligence or willful misconduct.

[81] See the Obama Administration's America's Gulf Coast: A Long Term Recovery Plan after the Deepwater Horizon Oil Spill (sometimes referred to as the "Mabus Report"), September 2010.

[82] See, e.g., Letter from Thomas Perrelli (U.S. Associate Attorney General) to Kenneth Feinberg (GCCF Administrator), September 17, 2010; and a subsequent letter from Perrelli to Feinberg, February 4, 2011.

[83] E.g., House Committee on Small Business (June 30, 2010); House Committee on the Judiciary (July 21, 2010); Senate Committee on Homeland Security and Governmental Affairs (July 22, 2010); and the House Committee on Energy and Commerce (July 27, 2010).

[84] The Senate Ad Hoc Subcommittee on Disaster Recovery of the Senate Committee on Homeland Security and Governmental Affairs held a hearing January 27, 2011. See also CRS Report R41684, *Oil Spill Legislation in the 112th Congress*, by Jonathan L. Ramseur.

[85] See, e.g., Letter from Congressman Scalise to Kenneth Feinberg, January 24, 2011; and Thomas Perrelli (U.S. Associate Attorney General) to Kenneth Feinberg (GCCF Administrator), February 4, 2011.

[86] See, e.g., Testimony of Lamar McKay (Chairman & President, BP America) before the Senate Committee on Energy and Natural Resources, May 11, 2010.

[87] Other motivations may have played a role, but that discussion is beyond the scope of this report.

[88] White House Press Release, June 16, 2010, at http://www.whitehouse.gov/the-press-office/fact-sheet-claims-andescrow. See also Deepwater Horizon Oil Spill Trust Agreement between BP Exploration and Production, Inc. as grantor of the funds, and individual and corporate trustees, dated Aug. 6, 2010.

[89] BP Press Release (August 23, 2010), at http://www.bp.com/genericarticle.do?categoryId=2012968&contentId= 7064597.

[90] The methodology is available at the GCCF website, http://www.gulfcoastclaimsfacility.com.

[91] In a February 19, 2011 interview with CBS News, Kenneth Feinberg stated that the assumption was based on "gathering together all the best expertise," reiterating that his assumption is a "reasonable estimate."

[92] Dr. John W. (Wes) Tunnell, Jr., (Harte Research Institute for Gulf of Mexico Studies, and Texas A&M University-Corpus Christi), "An Expert Opinion of when the Gulf of Mexico will return to pre-spill harvest status following the BP Deepwater Horizon MC 252 oil spill," January 31, 2011.

[93] http://www.gulfcoastclaimsfacility.com.

[94] See e.g., Senator Vitter's remarks at a hearing in the Senate Ad Hoc Subcommittee on Disaster Recovery in the Senate Committee on Homeland Security and Governmental Affairs, January 27, 2011. Data related to this argument were published in a story by David Hammer in the *Times-Picayune* January 26, 2011.

[95] In accordance with OPA, these would all be claims already filed with either BP or the GCCF. The claims would have been denied or the claimants were not satisfied with the outcome.

[96] OPA § 1012 (33 U.S.C. § 2712).

[97] See e.g., H.R. 6016 (Brady) and H.R. 3534 ("CLEAR Act," Section 707). Both bills passed the House in 2010.

[98] Bills that passed the House with a 45-day clock include H.R. 5629 (Oberstar) and H.R. 3534 ("CLEAR Act," Section 722). S. 3663 (Reid) would have shortened the wait-period to 30 days.

[99] CRS Report R41684, *Oil Spill Legislation in the 112th Congress*, by Jonathan L. Ramseur.

[100] See e.g., Letter from Bill McCollum (Attorney General of Florida) to Kenneth Feinberg (GCCF Administrator), August 20, 2010. See also Letter from Thomas Perrelli (Associate Attorney General) to Kenneth Feinberg (GCCF Administrator), February 4, 2011.

[101] GCCF, "Protocol for Emergency Advance Payments" (August 23, 2010).

[102] OPA § 1002; 33 U.S.C. § 2702 (emphasis added).

[103] National Commission on the BP Deepwater Horizon Oil Spill and Offshore Drilling, *Deep Water: The Gulf Oil Disaster and the Future of Offshore Drilling*, Report to the President, January 2011, p. 186.

[104] CRS Report R41684, *Oil Spill Legislation in the 112th Congress*, by Jonathan L. Ramseur.

[105] According to a recent Resources for the Future discussion paper, "the literature on oil offshore exploration and production, and fixed platforms specifically, remains relatively underdeveloped." The same paper included a "preliminary" analysis and found (among other things) "for an average platform (i.e., a platform with the sample's average age, annual production, number of producing wells, and other characteristics), each 100 feet of added depth increases the probability of a company-reported incident by 8.5 percent." See Lucija Muehlenbachs *et al.*, "Preliminary Empirical Assessment of Offshore Production Platforms in the Gulf of Mexico," Resources for the Future Discussion Paper, January 2011.

[106] See CRS Report RL33558, *Nuclear Energy Policy*, by Mark Holt.

[107] CRS Report R41684, *Oil Spill Legislation in the 112th Congress*, by Jonathan L. Ramseur.

[108] For example, the U.S. Coast Guard and the Bureau of Ocean Energy Management, Regulation, and Enforcement (BOEM) are conducting a joint investigation into the *Deepwater Horizon* incident to develop conclusions and recommendations as they relate to the explosion and loss of life on April 20, 2010. The final report is due no later than July 27, 2011. For more details, see http://www.deepwaterinvestigation.com.

In: Gulf Oil Spill of 2010...
Editors: C. R. Walsh, J. P. Duncan

ISBN: 978-1-61324-729-7
© 2012 Nova Science Publishers, Inc.

Chapter 2

OIL POLLUTION ACT OF 1990 (OPA): LIABILITY OF RESPONSIBLE PARTIES[*]

James E. Nichols

SUMMARY

The Oil Pollution Act of 1990 (OPA) establishes a framework that addresses the liability of responsible parties in connection with the discharge of oil into the navigable waters of the United States, adjoining shorelines, or the exclusive economic zone. Among other provisions, OPA limits certain liabilities of a responsible party in connection with discharges of oil into such areas. The liability limitations established by OPA are currently the subject of significant congressional interest in the wake of the Deepwater Horizon oil spill in the Gulf of Mexico.

A responsible party is strictly liable for removal costs plus damages resulting from an oil spill incident under OPA. A responsible party's liability under OPA is confined to specific categories of damages. Pursuant to OPA, the total liability for damages in connection with an oil spill is limited based on the type of vessel or facility involved, and the amount of oil discharged. OPA provides limited defenses which, if applicable, allow a responsible party to discharge its liability to persons injured by a discharge of oil into the navigable waters of the United States, adjoining shorelines, or the exclusive economic zone.

---

[*] This is an edited, reformatted and augmented version of a Congressional Research Service publication, CRS Report for Congress R41266, from www.crs.gov, dated June 2, 2010.

Claims for removal costs and certain damages must, with limited exception, be presented directly to the responsible party. In the event that a claim for removal costs or certain damages is not paid by the responsible party within 90 days, a claimant may present such a claim directly to the Oil Spill Liability Trust Fund or file suit in court. OPA and its regulations establish procedures for recovering removal costs and damages against the Oil Spill Liability Trust Fund.

This report addresses liability under OPA for removal costs and damages, and the basic procedure for recovering removal costs and damages from the Oil Spill Liability Trust Fund in the event that the responsible party fails to settle such claims.

BACKGROUND

The Oil Pollution Act of 1990 (OPA) establishes a framework that addresses the liability of responsible parties in connection with the discharge of oil into the navigable waters of the United States, adjoining shorelines, or the exclusive economic zone.[1] Among other provisions, OPA limits certain liabilities of a responsible party in connection with discharges of oil into such areas.[2]

OPA replaced the liability limitations established in Clean Water Act section 311[3] with much higher ones and expanded the class of persons authorized to recover removal costs from the responsible party to "any person" who has incurred removal costs in connection with a discharge of oil into covered waters.[4] Under the Clean Water Act, only the federal government could recover removal costs from the responsible party. Additionally, OPA created categories of damages for which the responsible party would be liable, within specified limits, based on the type of vessel or facility involved in the oil spill incident.[5] The following sections of this report will discuss liability under OPA for removal costs and damages, and the basic procedure for recovering removal costs and damages from the Oil Spill Liability Trust Fund[6] in the event that the responsible party fails to pay such claims.

LIABILITY OF RESPONSIBLE PARTIES[7]

Under OPA, a responsible party is strictly and jointly and severally liable for removal costs plus damages in connection with a discharge of oil into covered waters.[8] A responsible party's liability for damages, however, is limited under OPA.[9] The Oil Pollution Act of 1990 also states that additional liability may be imposed upon a responsible party under state law.[10]

Responsible Party Liability for Removal Costs

A responsible party in an oil spill incident is liable for removal costs[11] under OPA:

> Notwithstanding any other provision or rule of law, and subject to the provisions of [OPA], each responsible party for a vessel or a facility from which oil is discharged, or which poses the substantial threat of a discharge of oil, into or upon the navigable waters or adjoining shorelines or the exclusive economic zone is liable for the *removal costs* and damages ... that result from such incident.[12]

Removal costs may be recovered from a responsible party by the United States, affected states and Indian tribes, and by any person, to the extent that such person has undertaken removal actions pursuant to the National Contingency Plan, mandated by Clean Water Act section 311.[13] In general, claims for removal costs must be presented first to a responsible party.[14] If the party to whom the claim is presented denies all liability for the claim, or if the claim is not settled by payment within 90 days after the claim was presented, the claimant may elect either to initiate an action in court[15] against the responsible party or to present the claim directly to the Oil Spill Liability Trust Fund.[16]

In limited situations, however, certain claims for removal costs may be presented initially to the Oil Spill Liability Trust Fund.[17] For example, if the President (acting through the Director of the National Pollution Funds Center) advertises or otherwise notifies claimants in writing, then such claimants may bypass the responsible party and present claims for removal costs directly to the Fund.[18] A responsible party may also present claims for removal costs directly to the Fund.[19] The Governor of an affected state may present a claim directly to the Fund for removal costs incurred by that state.[20] Finally, a "United States claimant" may present a claim for removal costs in the event that a foreign offshore unit has discharged oil causing damage for which the Fund is otherwise liable.[21]

In the event that a person presents a claim for removal costs to the Fund, in addition to the aforementioned procedural requirements the claimant must establish (1) that the actions taken were necessary to prevent, minimize, or mitigate the effects of the oil spill incident; (2) that the removal costs were incurred as a result of the actions taken to prevent, minimize, or mitigate the effects of the oil spill incident; and (3) that the actions taken were determined by the Federal On-Scene Coordinator to be consistent with the National Contingency Plan, or were otherwise directed by the Federal On-Scene Coordinator.[22] The amount of compensation for removal cost claims against the Fund is "the total of uncompensated reasonable removal costs of actions taken that were determined by the [Federal On-Scene Coordinator] to be consistent with the National Contingency Plan or were directed by the [Federal On-Scene Coordinator]."[23]

Removal activities for which costs are being claimed must be coordinated by the Federal On-Scene Coordinator, except in "exceptional circumstances."[24] Accordingly, costs incurred for removal activities that are not coordinated with, or directed by, the Federal On-Scene Coordinator may not be recoverable against the Fund.[25]

Responsible Party Liability for Damages

Under OPA, responsible parties are liable for certain damages resulting from an oil spill incident.[26]

> Notwithstanding any other provision or rule of law, and subject to the provisions of [OPA], each responsible party for a vessel or a facility from which oil is discharged, or which poses the substantial threat of a discharge of oil, into or upon the navigable waters or adjoining shorelines or the exclusive economic zone is liable for the removal costs and *damages* ... that result from such incident.[27]

Generally, claims for damages must be presented first to a responsible party.[28] If the party to whom the claim is presented denies all liability for the claim, or if the claim is not settled by payment within 90 days after which the claim was presented, the claimant may elect either to initiate an action in court against the responsible party, or to present the claim directly to the Oil Spill Liability Trust Fund.[29]

As discussed below, some categories of damages are available for any person affected by the oil spill incident, while other categories of damages are only recoverable by the United States, states, and/or political subdivisions of states. Unlike removal costs, which are also determined on a per-incident and per-responsible party basis but are uncapped, damages are capped under OPA on a per-incident and per-responsible party basis, unless certain exceptions apply.[30] In the event that an exception applies to a particular oil spill incident, a responsible party's liability for damages under OPA is unlimited.[31] Although the act does not expressly prohibit the recovery of punitive damages against the responsible party, courts that have considered the issue have held that punitive damages are not recoverable against a responsible party under the act.[32]

Limits on Liability

Limitations on liability for damages under OPA are determined on a per-responsible party and per-incident basis.[33] The act further limits a responsible party's liability for damages based on the type of vessel or facility from which the discharge of oil flows. Specifically, the act provides different liability limits for (1) tank vessels; (2) vessels, generally; (3) offshore facilities (other than deepwater ports); (4) onshore facilities and deepwater ports; and (5) mobile offshore drilling units.[34]

For example, the Deepwater Horizon rig, which exploded in the Gulf of Mexico on April 20, 2010, and sank two days later, is classified as a mobile offshore drilling unit.[35] Initially, a mobile offshore drilling unit is deemed to be a tank vessel for the purposes of determining the responsible party's liability.[36] In the event that damages exceed the liability limits for discharges from tank vessels, the mobile offshore drilling unit is deemed to be an offshore facility.[37] Thus, if damages in the Gulf of Mexico exceed the liability limit established for tank vessels under OPA,[38] the Deepwater Horizon would be deemed an offshore facility, and the liability for the "responsible party" could be limited to removal costs plus $75 million.[39]

The limitations on liability mentioned above, however, are subject to several exceptions. Specifically, if the Deepwater Horizon oil spill incident is determined to be the result of gross negligence, willful misconduct, or violation of federal safety, construction, or operating regulations by a responsible party, then the relevant liability limitation is not applicable.[40] Additionally, the liability limitations established by the act are not applicable as to the responsible party if the responsible party fails or refuses to (a) report the incident (as required by law), (b) cooperate with federal removal activities, or (c) comply with the National Contingency Plan.[41]

Natural Resources

Responsible parties are liable to the United States, states, Indian tribes, or foreign governments for the harm to natural resources caused by an oil spill incident under OPA.[42] The term "natural resources" is broadly defined by the act to include "land, fish, wildlife, biota, air, water, ground water, drinking water supplies, and other such resources belonging to, managed by, held in trust by, appertaining to, or otherwise controlled by" the United States (including the resources of the exclusive economic zone), any state or local government, any Indian tribe, or any foreign government.[43]

Although the term "natural resources" is defined broadly, damages for "injury to, destruction of, loss of, or loss of use of, natural resources, including the reasonable costs of assessing the damages" are only recoverable by trustees acting on behalf of the United States, states, Indian tribes, or foreign governments.[44]

Generally, claims for natural resource damages must be presented directly to the responsible party. If not paid by the responsible party, OPA and its implementing regulations establish a procedure through which uncompensated natural resource damages may be recovered by affected claimants from the Oil Spill Liability Trust Fund.[45]

Real or Personal Property

Responsible parties are liable to claimants under OPA for harm to real or personal property resulting from an oil spill incident.[46] Specifically, a claimant who owns or leases an affected property may recover damages from the responsible party "for injury to, or economic losses resulting from destruction of, real or personal property."[47]

In the event that the responsible party fails to settle a claim for damages to real or personal property within 90 days, a claimant may present such a claim directly to the Oil Spill Liability Trust Fund. To recover damages against the Fund for "injury" to real or personal property, a claimant must prove (1) property ownership or control; (2) property damage or injury; (3) the cost of repairing or replacing the property; and (4) the property values before and after the oil spill incident.[48]

To recover damages against the Fund for "economic loss resulting from the destruction" of real or personal property, in addition to the aforementioned requirements listed above, a claimant must prove (1) that the affected property was not available for use, and if it had been available for use, the value of that use; (2) the availability of substitute property, and if used in lieu of the injured property, the costs associated with such use; and (3) a nexus between the economic losses claimed and the injury to the property.[49]

Loss of Subsistence Use

Under OPA, responsible parties are liable to claimants for the "loss of subsistence use of natural resources."[50] Any claimant who "uses natural resources [for subsistence purposes] which have been injured, destroyed, or lost" as a result of an oil spill incident may recover damages from the responsible party.[51]

In the event that a responsible party fails to settle a claim for damages for loss of subsistence use of natural resources within 90 days, a claimant may present such a claim directly to the Oil Spill Liability Trust Fund. To recover against the Fund, a claimant must (1) identify each specific natural resource for which compensation for loss of subsistence use is claimed; (2) describe the actual subsistence use made of each natural resource; (3) describe how and to what extent the subsistence use was affected by the oil spill incident; (4) describe the mitigation efforts undertaken by the claimant; and (5) describe alternative sources or means of subsistence available to the claimant during the period of time for which loss of subsistence is claimed, and any compensation available to the claimant for loss of subsistence.

Government Revenues

Under OPA, responsible parties are liable for damages "equal to the net loss of taxes, royalties, rents, fees, or net profit shares due to the injury, destruction, or loss of real property, personal property, or natural resources."[52] Only the United States, states, and political subdivisions of states may recover damages from the responsible party for lost government revenues.[53] Accordingly, private entities may not recover damages, either against the responsible party or against the Oil Spill Liability Trust Fund, under this section of the act.

In the event that a responsible party fails to settle a claim for damages for lost government revenues within 90 days, a claimant may present such a claim directly to the Oil Spill Liability Trust Fund. To recover against the Fund, a claimant must (1) identify and describe the economic loss for which compensation is claimed; (2) prove a causal nexus between the loss of revenue and the destruction of real or personal property, or natural resources; (3) provide the total assessment or revenue collected for comparable revenue periods; and (4) establish the net loss of revenue.[54]

The total compensation allowable for lost government revenue claims presented directly to the Fund is the total net revenue actually lost as a result of the oil spill incident.[55]

Profits and Earning Capacity

Damages "equal to the loss of profits or impairment of earning capacity due to the injury, destruction, or loss of real property, personal property, or natural resources" are recoverable by any claimant against the responsible party under OPA.[56] As mentioned above, the act broadly defines the term "natural resources" to include land, fish, wildlife, biota, air, water, ground water, drinking water supplies, and "other such resources."[57]

In the event that a responsible party fails to settle a claim for damages for lost profits and earning capacity within 90 days, a claimant may present such a claim directly to the Oil Spill Liability Trust Fund. To recover against the Fund, a claimant must prove (1) that real or personal property or natural resources have been injured, destroyed, or lost; (2) that the claimant's income was reduced as a consequence of the injury, and the amount of the reduction; (3) the amount of profits or earnings in comparable periods as compared to the period when the claimed loss was suffered; and (4) whether alternative employment or business was available and undertaken.[58]

The amount of compensation allowable for "lost profits and earning capacity" claims presented directly to the Fund is limited to the actual net reduction or loss of earnings or profits suffered.[59]

Public Services

In the event that an affected state or political subdivision of an affected state incurs costs related to the provision of "increased or additional public services during or after removal activities, including protection from fire, safety, or health hazards, caused by a discharge of oil," the responsible party is liable for such costs.[60] Under OPA, only states and their political subdivisions may recover the costs associated with providing increased or additional public services (to the extent that such states are affected by the discharge of oil).[61] Accordingly, the United States is precluded from recovering costs related to the provision of public services from the responsible party under the act.

In the event that a responsible party fails to settle a claim for damages related to the provision of increased or additional public services within 90 days, a claimant may present such a claim directly to the Oil Spill Liability Trust Fund. To recover against the Fund, a claimant must establish (1) the specific type of the public services provided and the need for such services; (2) that the public services occurred during removal activities; (3) that the services were provided as a result of the oil spill incident and would have not otherwise been provided; and (4) the total costs for the provision of public services and the methods used to compute such costs.[62]

The total amount of compensation allowable for claims presented directly to the Fund for the increased or additional provision of public services is the net cost of the increased or additional service provided by the affected state or political subdivision of the affected state.[63]

Defenses to Liability

OPA provides limited defenses to liability. A responsible party is not liable for removal costs or the damages mentioned above if it can establish—by a preponderance of the evidence—that the discharge or substantial threat of discharge of oil and the resulting damages or removal costs were caused solely by

> (1) an act of God; (2) an act of war; (3) an act or omission of a third party, other than an employee or agent of the responsible party or a third party whose act or omission occurs in connection with any contractual relationship with the responsible party (except where the sole contractual arrangement arises in connection with carriage by a common carrier by rail), if the responsible party establishes, by a preponderance of the evidence, that the responsible party—(A) exercised due care with respect to the oil concerned, taking into consideration the characteristics of the oil and in light of all relevant facts and circumstances; and (B) took precautions against foreseeable acts or omissions of any such third party and the foreseeable consequences of those acts or omissions; or (4) any combination of paragraphs (1), (2), (3).[64]

In other words, OPA requires a responsible party to pay for removal costs plus certain damages resulting from the oil spill, unless the incident was caused by a serious and unanticipated naturally occurring event (e.g., earthquake, hurricane, tornado, etc.), an act of war, an act or omission by a third party with no contractual relationship to the responsible party, or any combination of the aforementioned circumstances. For the "act or omission of a third party" defense to be available to the responsible party, the responsible party must have been operating its facility or vessel in a non-negligent manner.[65]

Pursuant to OPA, these defenses to liability are not available to a responsible party who fails or refuses (a) to report the oil spill incident as required by law, (b) to provide reasonable cooperation and assistance with removal activities, or (c) to comply, without sufficient cause, with the

President's general removal authority.[66] Additionally, if a responsible party has actual knowledge of a discharge or a substantial threat of a discharge of oil, it cannot escape liability for removal costs or damages by transferring ownership of the facility (or property upon which the facility is located) to an innocent third party.[67]

OIL SPILL LIABILITY TRUST FUND

Although the Oil Spill Liability Trust Fund is not the focus of this report, the purpose and operation of the Fund should be briefly explained. The Oil Spill Liability Trust Fund is a federally administered trust fund that may be used to pay costs related to federal and state oil spill removal activities; costs incurred by federal, state, and Indian tribe trustees for natural resource damage assessments; and unpaid damages claims.[68] The Fund is financed by a per-barrel tax on crude oil received at United States refineries, and on petroleum products imported into the United States for consumption.[69] The maximum amount of money that may be withdrawn from the Fund is $1 billion per incident.[70] Currently, the Fund may not receive advances from the United States Treasury, as its authority to borrow expired December 31, 1994.[71] The United States Attorney General, however, may commence an action on behalf of the Fund, against a responsible party, to recover any money paid by the Fund to any claimant pursuant to OPA.[72]

APPENDIX A. OIL POLLUTION ACT OF 1990: DEFINITIONS

The Oil Pollution Act of 1990 defines the key terms used throughout the act. For the reader's convenience, this appendix provides definitions of key terms as they appear in 33 U.S.C. § 2701.

Definitions

For the purposes of this Act, the term—

(1) "**act of God**" means an unanticipated grave natural disaster or other natural phenomenon of an exceptional, inevitable, and

irresistible character the effects of which could not have been prevented or avoided by the exercise of due care or foresight;

(2) **"barrel"** means 42 United States gallons at 60 degrees Fahrenheit;

(3) **"claim"** means a request, made in writing for a sum certain, for compensation for damages or removal costs resulting from an incident;

(4) **"claimant"** means any person or government who presents a claim for compensation under this title;

(5) **"damages"** means damages specified in section 1002(b) of this Act [33 U.S.C. § 2702(b)], and includes the cost of assessing these damages;

(6) **"deepwater port"** is a facility licensed under the Deepwater Port Act of 1974 (33 U.S.C. § 1501-1524);

(7) **"discharge"** means any emission (other than natural seepage), intentional or unintentional, and includes, but is not limited to, spilling, leaking, pumping, pouring, emitting, emptying, or dumping;

(8) **"exclusive economic zone"** means the zone established by Presidential Proclamation Numbered 5030, dated March 10, 1983, including the ocean waters of the areas referred to as "eastern special areas" in Article 3(1) of the Agreement between the United States of America and the Union of Soviet Socialist Republics on the Maritime Boundary, signed June 1, 1990;

(9) **"facility"** means any structure, group of structures, equipment, or device (other than a vessel) which is used for one or more of the following purposes: exploring for, drilling for, producing, storing, handling, transferring, processing, or transporting oil. This term includes any motor vehicle, rolling stock, or pipeline used for one or more of these purposes;

(10) **"foreign offshore unit"** means a facility which is located, in whole or in part, in the territorial sea or on the continental shelf of a foreign country and which is or was used for one or more of the following purposes: exploring for, drilling for, producing, storing, handling, transferring, processing, or transporting oil produced from the seabed beneath the foreign country's territorial sea or from the foreign country's continental shelf;

(11) **"Fund"** means the Oil Spill Liability Trust Fund, established by section 9509 of the Internal Revenue Code of 1986 (26 U.S.C. § 9509);

(12) **"gross ton"** has the meaning given that term by the Secretary under part J of title 46, United States Code [46 U.S.C. §§ 14101 *et seq.*];

(13) **"guarantor"** means any person, other than the responsible party, who provides evidence of financial responsibility for a responsible party under this Act;

(14) **"incident"** means any occurrence or series of occurrences having the same origin, involving one or more vessels, facilities, or any combination thereof, resulting in the discharge or substantial threat of discharge of oil;

(15) **"Indian tribe"** means any Indian tribe, band, nation, or other organized group or community, but not including any Alaska Native regional or village corporation, which is recognized as eligible for the special programs and services provided by the United States to Indians because of their status as Indians and has governmental authority over lands belonging to or controlled by the tribe;

(16) **"lessee"** means a person holding a leasehold interest in an oil or gas lease on lands beneath navigable waters (as that term is defined in section 2(a) of the Submerged Lands Act (43 U.S.C. § 1301(a))) or on submerged lands of the Outer Continental Shelf, granted or maintained under applicable State law or the Outer Continental Shelf Lands Act (43 U.S.C. § 1331 *et seq.*);

(17) **"liable"** or **"liability"** shall be construed to be the standard of liability which obtains under section 311 of the Federal Water Pollution Control Act (33 U.S.C. § 1321);

(18) **"mobile offshore drilling unit"** means a vessel (other than a self-elevating lift vessel) capable of use as an offshore facility;

(19) **"National Contingency Plan"** means the National Contingency Plan prepared and published under section 311(d) of the Federal Water Pollution Control Act [33 USCS § 1321(d)], as amended by this Act, or revised under section 105 of the Comprehensive Environmental Response, Compensation, and Liability Act (42 U.S.C. § 9605);

(20) **"natural resources"** includes land, fish, wildlife, biota, air, water, ground water, drinking water supplies, and other such resources belonging to, managed by, held in trust by, appertaining to, or otherwise controlled by the United States (including the resources of the exclusive economic zone), any State or local government or Indian tribe, or any foreign government;

(21) **"navigable waters"** means the waters of the United States, including the territorial sea;

(22) **"offshore facility"** means any facility of any kind located in, on, or under any of the navigable waters of the United States, and any facility of any kind which is subject to the jurisdiction of the United States and is located in, on, or under any other waters, other than a vessel or a public vessel;

(23) **"oil"** means oil of any kind or in any form, including petroleum, fuel oil, sludge, oil refuse, and oil mixed with wastes other than dredged spoil, but does not include any substance which is specifically listed or designated as a hazardous substance under subparagraphs (A) through (F) of section 101(14) of the Comprehensive Environmental Response, Compensation, and Liability Act (42 U.S.C. § 9601) and which is subject to the provisions of that Act;

(24) **"onshore facility"** means any facility (including, but not limited to, motor vehicles and rolling stock) of any kind located in, on, or under, any land within the United States other than submerged land;

(25) the term **"Outer Continental Shelf facility"** means an offshore facility which is located, in whole or in part, on the Outer Continental Shelf and is or was used for one or more of the following purposes: exploring for, drilling for, producing, storing, handling, transferring, processing, or transporting oil produced from the Outer Continental Shelf;

(26) **"owner or operator"**—

(A) means—

(i) in the case of a vessel, any person owning, operating, or chartering by demise, the vessel;

(ii) in the case of an onshore or offshore facility, any person owning or operating such facility;

(iii) in the case of any abandoned offshore facility, the person who owned or operated such facility immediately prior to such abandonment;

(iv) in the case of any facility, title or control of which was conveyed due to bankruptcy, foreclosure, tax delinquency, abandonment, or similar means to a unit of State or local government, any person who owned, operated, or otherwise controlled activities at such facility immediately beforehand;

(v) notwithstanding subparagraph (B)(i), and in the same manner and to the same extent, both procedurally and substantively, as any nongovernmental entity, including for purposes of liability under section 1002, any State or local government that has caused or contributed to a discharge or substantial threat of a discharge of oil from a vessel or facility ownership or control of which was acquired involuntarily through—

> I. seizure or otherwise in connection with law enforcement activity;
> II. bankruptcy;
> III. tax delinquency;
> IV. abandonment; or
> V. other circumstances in which the government involuntarily acquires title by virtue of its function as sovereign;

(vi) notwithstanding subparagraph (B)(ii), a person that is a lender and that holds indicia of ownership primarily to protect a security interest in a vessel or facility if, while the borrower is still in possession of the vessel or facility encumbered by the security interest, the person—

(I) exercises decision making control over the environmental compliance related to the vessel or facility, such that the person has undertaken responsibility for oil handling or disposal practices related to the vessel or facility; or

(II) exercises control at a level comparable to that of a manager of the vessel or facility, such that the person has assumed or manifested responsibility—

(aa) for the overall management of the vessel or facility encompassing day-to-day decision making with respect to environmental compliance; or

(bb) over all or substantially all of the operational functions (as distinguished from financial or administrative functions) of the vessel or facility other than the function of environmental compliance; and

(B) does not include—
(i) A unit of state or local government that acquired ownership or control of a vessel or facility involuntarily through—

    I.   seizure or otherwise in connection with law enforcement activity;
    II.   bankruptcy;
    III.   tax delinquency;
    IV.   abandonment; or
    V.   other circumstances in which the government involuntarily acquires title by virtue of its function as sovereign;

(ii) a person that is a lender that does not participate in management of a vessel or facility, but holds indicia of ownership primarily to protect the security interest of the person in the vessel or facility; or

(iii) a person that is a lender that did not participate in management of a vessel or facility prior to foreclosure, notwithstanding that the person—

    I.   forecloses on the vessel or facility; and
    II.   after foreclosure, sells, re-leases (in the case of a lease finance transaction), or liquidates the vessel or facility, maintains business activities, winds up operations, undertakes a removal action under section 311(c) of the Federal Water Pollution Control Act (33 U.S.C. § 1321(c)) or under the direction of an on-scene coordinator appointed under the National Contingency Plan, with respect to the vessel or facility, or takes any other measure to preserve, protect, or prepare the vessel or facility prior to sale or disposition, if the person seeks to sell, re-lease (in the case of a lease finance transaction), or otherwise divest the person of the vessel or facility at the earliest practicable, commercially reasonable time, on commercially reasonable terms, taking into account market conditions and legal and regulatory requirements;

(27) "**person**" means an individual, corporation, partnership, association, State, municipality, commission, or political subdivision of a State, or any interstate body;

(28) "**permittee**" means a person holding an authorization, license, or permit for geological exploration issued under section 11 of the Outer Continental Shelf Lands Act (43 U.S.C. § 1340) or applicable State law;

(29) **"public vessel"** means a vessel owned or bareboat chartered and operated by the United States, or by a State or political subdivision thereof, or by a foreign nation, except when the vessel is engaged in commerce;

(30) **"remove"** or **"removal"** means containment and removal of oil or a hazardous substance from water and shorelines or the taking of other actions as may be necessary to minimize or mitigate damage to the public health or welfare, including, but not limited to, fish, shellfish, wildlife, and public and private property, shorelines, and beaches;

(31) **"removal costs"** means the costs of removal that are incurred after a discharge of oil has occurred or, in any case in which there is a substantial threat of a discharge of oil, the costs to prevent, minimize, or mitigate oil pollution from such an incident;

(32) **"responsible party"** means the following:

- a) Vessels. In the case of a vessel, any person owning, operating, or demise chartering the vessel.
- b) Onshore facilities. In the case of an onshore facility (other than a pipeline), any person owning or operating the facility, except a Federal agency, State, municipality, commission, or political subdivision of a State, or any interstate body, that as the owner transfers possession and right to use the property to another person by lease, assignment, or permit.
- c) Offshore facilities. In the case of an offshore facility (other than a pipeline or a deepwater port licensed under the Deepwater Port Act of 1974 (33 U.S.C. § 1501 *et seq.*)), the lessee or permittee of the area in which the facility is located or the holder of a right of use and easement granted under applicable State law or the Outer Continental Shelf Lands Act (43 U.S.C. §§ 1301- 1356) for the area in which the facility is located (if the holder is a different person than the lessee or permittee), except a Federal agency, State, municipality, commission, or political subdivision of a State, or any interstate body, that as owner transfers possession and right to use the property to another person by lease, assignment, or permit.
- d) Deepwater ports. In the case of a deepwater port licensed under the Deepwater Port Act of 1974 (33 U.S.C. §§ 1501-1524), the licensee.
- e) Pipelines. In the case of a pipeline, any person owning or operating the pipeline.

f) Abandonment. In the case of an abandoned vessel, onshore facility, deepwater port, pipeline, or offshore facility, the persons who would have been responsible parties immediately prior to the abandonment of the vessel or facility.

(33) "**Secretary**" means the Secretary of the department in which the Coast Guard is operating;

(34) "**tank vessel**" means a vessel that is constructed or adapted to carry, or that carries, oil or hazardous material in bulk as cargo or cargo residue, and that—

a) is a vessel of the United States;
b) operates on the navigable waters; or
c) transfers oil or hazardous material in a place subject to the jurisdiction of the United States;

(35) "**territorial seas**" means the belt of the seas measured from the line of ordinary low water along that portion of the coast which is in direct contact with the open sea and the line marking the seaward limit of inland waters, and extending seaward a distance of 3 miles;

(36) "**United States**" and "**State**" mean the several States of the United States, the District of Columbia, the Commonwealth of Puerto Rico, Guam, American Samoa, the United States Virgin Islands, the Commonwealth of the Northern Marianas, and any other territory or possession of the United States;

(37) "**vessel**" means every description of watercraft or other artificial contrivance used, or capable of being used, as a means of transportation on water, other than a public vessel;

(38) "**participate in management**"—

A. means actually participating in the management or operational affairs of a vessel or facility; and
B. does not include merely having the capacity to influence, or the unexercised right to control, vessel or facility operations; and
C. does not include—
I. performing an act or failing to act prior to the time at which a security interest is created in a vessel or facility;
II. holding a security interest or abandoning or releasing a security interest;

III. including in the terms of an extension of credit, or in a contract or security agreement relating to the extension, a covenant, warranty, or other term or condition that relates to environmental compliance;
IV. monitoring or enforcing the terms and conditions of the extension of credit or security interest;
V. monitoring or undertaking one or more inspections of the vessel or facility;
VI. requiring a removal action or other lawful means of addressing a discharge or substantial threat of a discharge of oil in connection with the vessel or facility prior to, during, or on the expiration of the term of the extension of credit;
VII. providing financial or other advice or counseling in an effort to mitigate, prevent, or cure default or diminution in the value of the vessel or facility;
VIII. restructuring, renegotiating, or otherwise agreeing to alter the terms and conditions of the extension of credit or security interest, exercising forbearance;
IX. exercising other remedies that may be available under applicable law for the breach of a term or condition of the extension of credit or security agreement; or
X. conducting a removal action under 311(c) of the Federal Water Pollution Control Act (33 U.S.C. § 1321(c)) or under the direction of an on-scene coordinator appointed under the National Contingency Plan, if such actions do not rise to the level of participating in management under subparagraph (A) of this paragraph and paragraph (26)(A)(vi);

(39) "**extension of credit**" has the meaning provided in section 101(20)(G)(i) of the Comprehensive Environmental Response, Compensation and Liability Act of 1980 (42 U.S.C. § 9601(20)(G)(i));

(40) "**financial or administrative function**" has the meaning provided in section 101(20)(G)(ii) of the Comprehensive Environmental Response, Compensation and Liability Act of 1980 (42 U.S.C. § 9601(20)(G)(ii));

(41) "**foreclosure**" and "**foreclose**" each has the meaning provided in section 101(20)(G)(iii) of the Comprehensive Environmental Response, Compensation and Liability Act of 1980 (42 U.S.C. § 9601(20)(G)(iii));

(42) "**lender**" has the meaning provided in section 101(20)(G)(iv) of the Comprehensive Environmental Response, Compensation and Liability Act of 1980 (42 U.S.C. § 9601(20)(G)(iv));

(43) **"operational function"** has the meaning provided in section 101(20)(G)(v) of the Comprehensive Environmental Response, Compensation and Liability Act of 1980 (42 U.S.C. § 9601(20)(G)(v)); and

(44) **"security interest"** has the meaning provided in section 101(20)(G)(vi) of the Comprehensive Environmental Response, Compensation and Liability Act of 1980 (42 U.S.C. § 9601(20)(G)(vi)).

APPENDIX B. OIL POLLUTION ACT OF 1990: LIMITATIONS ON LIABILITY

The Oil Pollution Act of 1990 establishes limitations on the liability of responsible parties for certain damages caused by a discharge of oil into the navigable waters of the United States. For the reader's convenience, this appendix provides the liability limitation language as it appears in 33 U.S.C. § 2704.

Limits on liability

(a) General rule. Except as otherwise provided in this section, the total of the liability of a responsible party under section 1002 [33 USCS § 2702] and any removal costs incurred by, or on behalf of, the responsible party, with respect to each incident shall not exceed—
(1) for a tank vessel, the greater of—

  A. with respect to a single-hull vessel, including a single-hull vessel fitted with double sides only or a double bottom only, $3,000 per gross ton;
  B. with respect to a vessel other than a vessel referred to in subparagraph (A), $1,900 per gross ton; or
  C. (i) with respect to a vessel greater than 3,000 gross tons that is—

(I) a vessel described in subparagraph (A), $22,000,000; or
(II) a vessel described in subparagraph (B), $16,000,000; or
(ii) with respect to a vessel of 3,000 gross tons or less that is—
(I) a vessel described in subparagraph (A), $6,000,000; or
(II) a vessel described in subparagraph (B), $4,000,000;

(2) for any other vessel, $950 per gross ton or $800,000, whichever is greater;

(3) for an offshore facility except a deepwater port, the total of all removal costs plus $75,000,000; and

(4) for any onshore facility and a deepwater port, $350,000,000.

(b) Division of liability for mobile offshore drilling units.

(1) Treated first as tank vessel. For purposes of determining the responsible party and applying this Act and except as provided in paragraph (2), a mobile offshore drilling unit which is being used as an offshore facility is deemed to be a tank vessel with respect to the discharge, or the substantial threat of a discharge, of oil on or above the surface of the water.

(2) Treated as facility for excess liability. To the extent that removal costs and damages from any incident described in paragraph (1) exceed the amount for which a responsible party is liable (as that amount may be limited under subsection (a)(1)), the mobile offshore drilling unit is deemed to be an offshore facility. For purposes of applying subsection (a)(3), the amount specified in that subsection shall be reduced by the amount for which the responsible party is liable under paragraph (1).

(c) Exceptions.

(1) Acts of responsible party. Subsection (a) does not apply if the incident was proximately caused by—

A. gross negligence or willful misconduct of, or

B. the violation of an applicable Federal safety, construction, or operating regulation by, the responsible party, an agent or employee of the responsible party, or a person acting pursuant to a contractual relationship with the responsible party (except where the sole contractual arrangement arises in connection with carriage by a common carrier by rail).

(2) Failure or refusal of responsible party. Subsection (a) does not apply if the responsible party fails or refuses—

A. to report the incident as required by law and the responsible party knows or has reason to know of the incident;

B. to provide all reasonable cooperation and assistance requested by a responsible official in connection with removal activities; or

C. without sufficient cause, to comply with an order issued under subsection (c) or (e) of section 311 of the Federal Water Pollution

Control Act (33 U.S.C. § 1321), as amended by this Act, or the Intervention on the High Seas Act (33 U.S.C. § 1471 *et seq.*).

(3) OCS facility or vessel. Notwithstanding the limitations established under subsection (a) and the defenses of section 1003 [33 U.S.C. § 2703], all removal costs incurred by the United States Government or any State or local official or agency in connection with a discharge or substantial threat of a discharge of oil from any Outer Continental Shelf facility or a vessel carrying oil as cargo from such a facility shall be borne by the owner or operator of such facility or vessel.

(4) Certain tank vessels. Subsection (a)(1) shall not apply to—

a tank vessel on which the only oil carried as cargo is an animal fat or vegetable oil, as those terms are used in section 2 of the Edible Oil Regulatory Reform Act [33 U.S.C. § 2720]; and

a tank vessel that is designated in its certificate of inspection as an oil spill response vessel (as that term is defined in section 2101 of title 46, United States Code) and that is used solely for removal.

(d) Adjusting limits of liability.

(1) Onshore facilities. Subject to paragraph (2), the President may establish by regulation, with respect to any class or category of onshore facility, a limit of liability under this section of less than $350,000,000, but not less than $8,000,000, taking into account size, storage capacity, oil throughput, proximity to sensitive areas, type of oil handled, history of discharges, and other factors relevant to risks posed by the class or category of facility.

(2) Deepwater ports and associated vessels.

   A. Study. The Secretary shall conduct a study of the relative operational and environmental risks posed by the transportation of oil by vessel to deepwater ports (as defined in section 3 of the Deepwater Port Act of 1974 (33 U.S.C. § 1502)) versus the transportation of oil by vessel to other ports. The study shall include a review and analysis of offshore lightering practices used in connection with that transportation, an analysis of the volume of oil transported by vessel using those practices, and an analysis of the frequency and volume of oil discharges which occur in connection with the use of those practices.

   B. Report. Not later than 1 year after the date of the enactment of this Act [enacted Aug. 18, 1990], the Secretary shall submit to the Congress a report on the results of the study conducted under subparagraph (A).

C. Rulemaking proceeding. If the Secretary determines, based on the results of the study conducted under [this] subparagraph (A), that the use of deepwater ports in connection with the transportation of oil by vessel results in a lower operational or environmental risk than the use of other ports, the Secretary shall initiate, not later than the 180th day following the date of submission of the report to the Congress under subparagraph (B), a rulemaking proceeding to lower the limits of liability under this section for deepwater ports as the Secretary determines appropriate. The Secretary may establish a limit of liability of less than $350,000,000, but not less than $50,000,000, in accordance with paragraph (1).

(3) Periodic reports. The President shall, within 6 months after the date of the enactment of this Act [enacted Aug. 18, 1990], and from time to time thereafter, report to the Congress on the desirability of adjusting the limits of liability specified in subsection (a).

(4) Adjustment to reflect Consumer Price Index. The President, by regulations issued not later than 3 years after the date of enactment of the Delaware River Protection Act of 2006 [enacted July 11, 2006] and not less than every 3 years thereafter, shall adjust the limits on liability specified in subsection (a) to reflect significant increases in the Consumer Price Index.

End Notes

[1] 33 U.S.C. § 2701 et seq. Under OPA, the term "navigable waters" means "the waters of the United States, including the territorial sea." 33 U.S.C. § 2701(21).

[2] A recently introduced bill in the Senate, S. 3305, would raise OPA's limitation on liability for offshore facilities from $75 million to $10 billion, if enacted. The House companion bill to S. 3305, H.R. 5214, would amend OPA's limitation on liability for offshore facilities in the same manner as S. 3305, if enacted.

[3] 33 U.S.C. § 1321.

[4] 33 U.S.C. § 2702(b)(1)(B).

[5] 33 U.S.C. § 2702(b)(2)(A)-(F).

[6] The Oil Spill Liability Trust Fund is a federally administered trust fund that is financed by a per-barrel tax on petroleum products produced for consumption within the United States. 26 U.S.C. §§ 4611.

[7] Under OPA, the term "responsible party" refers to the owner or operator of a vessel or facility from which oil is discharged. 33 U.S.C § 2701(32). See Appendix A for OPA's definition of the term "responsible party."

[8] Under OPA, the terms "liable" and "liability" are "construed to be the standard of liability which obtains under section 311 of the [Clean Water Act]." Courts have interpreted section

311 of the Clean Water Act as imposing strict liability on parties responsible for the discharge of oil or hazardous substances into the waters of the United States. *See* United States v. New York, 481 F. Supp. 4 (D.N.Y. 1979).

[9] Unlike removal costs, which are uncapped, the responsible party's liability for damages under OPA is limited based on the type of vessel or facility involved, and the amount of oil discharged. 33 U.S.C. §§ 2704(a) and (b).

[10] 33 U.S.C. §§ 2718(a) and (c).

[11] Under OPA, the term "removal costs" means "the costs of removal that are incurred after a discharge of oil has occurred or, in any case in which there is a substantial threat of a discharge of oil, the costs to prevent, minimize, or mitigate oil pollution from such an incident." 33 U.S.C. § 2701(31).

[12] 33 U.S.C. § 2702(a) (emphasis supplied).

[13] 33 U.S.C. §§ 2702(b)(1)(A) and (B). The National Contingency Plan is authorized by Clean Water Act § 311(d). 33 U.S.C. § 1321(d). The implementing regulations promulgated by the Environmental Protection Agency are set forth at 40 CFR § 300.1 *et seq*.

[14] 33 U.S.C. § 2713(a). Under OPA, the term "claim" means "a request, made in writing for a sum certain, for compensation for damages or removal costs resulting from an [oil spill] incident." 33 U.S.C. § 2701(3).

[15] The Oil Pollution Act of 1990 provides federal district courts with "exclusive original jurisdiction over all controversies arising under [OPA]." 33 U.S.C. § 2717(b). State courts "of competent jurisdiction" may consider claims for damages or removal costs arising under OPA or applicable state law. 33 U.S.C. § 2717(c).

[16] 33 U.S.C. § 2713(c). Claims for removal costs must be presented within six (6) years after the date of completion of all removal activities related to the oil spill incident. 33 U.S.C. § 2712(h)(1). A guide for filing claims against the Oil Spill Liability Trust Fund can be found on the U.S. Coast Guard's National Pollution Fund Center website at http://www.uscg.mil/npfc/Claims/default.asp (last visited May 24, 2010).

[17] 33 U.S.C. § 2713(b).

[18] 33 U.S.C. § 2713(b)(1)(A); 33 CFR § 136.103(b)(1).

[19] 33 U.S.C. § 2708 (responsible party entitled to recover against Fund). *See* 33 U.S.C. § 2713(b)(1)(B); 33 CFR § 136.103(b)(2).

[20] 33 U.S.C. § 2713(b)(1)(C); 33 CFR § 136.103(b)(3).

[21] 33 U.S.C. § 2713(b)(1)(D); 33 CFR § 136.103(b)(4). The term "United States claimant" is not specifically defined by OPA or its implementing regulations.

[22] 33 U.S.C. § 2713(e); 33 CFR §§ 136.203(a), (b), and (c). The Federal On-Scene Coordinator of the Deepwater Horizon oil spill in the Gulf of Mexico is U.S. Coast Guard 8th District Deputy Commander, Rear Admiral James A. Watson.

[23] 33 U.S.C. § 2713(e); 33 CFR § 136.205.

[24] *Id*.

[25] Costs incurred for removal activities that are not coordinated with or directed by the Federal On-Scene Coordinator, however, may be presented by any claimant to the responsible party. 33 U.S.C. § 2713(a).

[26] Under OPA, the term "damages" means "damages specified in [33 U.S.C. § 2702(b)], *and includes the costs of assessing these damages*." 33 U.S.C. § 2701(5) (emphasis supplied). The standards and procedures for conducting natural resource damage assessments are set forth in regulations promulgated by the National Oceanic and Atmospheric Administration pursuant to OPA. 33 U.S.C. § 2706(e); 15 C.F.R. §§ 990.10 through 990.66.

[27] 33 U.S.C. § 2702(a) (emphasis supplied).

[28] 33 U.S.C. § 2713(a). Under OPA, the term "claim" means "a request, made in writing for a sum certain, for compensation for damages or removal costs resulting from an [oil spill] incident." 33 U.S.C. § 2701(3).

[29] 33 U.S.C. § 2713(c). Claims for the recovery of damages for an oil spill incident must be presented within three years of the oil spill incident. 33 U.S.C. § 2712(h)(2). A guide for filing claims against the Oil Spill Liability Trust Fund can be found on the U.S. Coast Guard's National Pollution Fund Center website at http://www.uscg.mil/npfc/ Claims/default.asp (last visited May 24, 2010).

[30] See 33 U.S.C. § 2704(c) (list of exceptions to liability limitations).

[31] Id.

[32] See South Port Marine, LLC v. Gulf Oil, LP, 234 F.3d 58 (1st Cir. 2000) (OPA displaces maritime-law punitive damages); Clausen v. M/V New Carissa, 171 F. Supp. 2d 1127 (D. Or. 2003) (OPA provides exclusive federal remedy for property damage claims resulting from oil spill, and thus precludes award of punitive damages for any claim for which the act could provide relief).

[33] 33 U.S.C. § 2704(a).

[34] Id. For definitions of the types of facilities covered by OPA, see Appendix A.

[35] See http://www.deepwaterinvestigation.com/go/doc/3043/558647/ (U.S. Coast Guard and Minerals Management Service joint press release referring to Deepwater Horizon oil rig as a mobile offshore drilling unit) (last visited on May 24, 2010).

[36] 33 U.S.C. § 2704(b)(1). The Federal On-Scene Coordinator designates the source of the discharge for the purposes of determining liability under OPA. 33 U.S.C § 2714(a); 33 C.F.R. § 136.305. The Federal On-Scene Coordinator of the ongoing oil spill in the Gulf of Mexico is U.S. Coast Guard 8th District Deputy Commander, Rear Admiral James A. Watson.

[37] 33 U.S.C. § 2704(b)(2).

[38] For the purposes of determining the liability limitation for a specific vessel or facility, OPA uses a formula that is based on the weight of the vessel or facility owned or operated by the responsible party. See Appendix B for OPA limitations on liability.

[39] 33 U.S.C. § 2704(b)(2).

[40] 33 U.S.C. §§ 2704(c)(1)(A) and (B).

[41] 33 U.S.C. §§ 2704(c)(2)(A), (B), and (C).

[42] 33 U.S.C. § 2702(b)(2)(A).

[43] 33 U.S.C. § 2701(20).

[44] 33 U.S.C. § 2702(b)(2)(A).

[45] 33 U.S.C. § 2713(e); 33 CFR Part 136, subparts B and C.

[46] 33 U.S.C. § 2702(b)(2)(B).

[47] 33 U.S.C. § 2702(b)(2)(B).

[48] 33 U.S.C. § 2713(e); 33 CFR §§ 136.215(a)(1)-(4).

[49] 33 U.S.C. § 2713(e); 33 CFR §§ 136.215(b)(1)-(3).

[50] 33 U.S.C. § 2702(b)(2)(C).

[51] Id.

[52] 33 U.S.C. § 2702(b)(2)(D).

[53] Id.

[54] 33 U.S.C. § 2713(e); 33 CFR §§ 136.227(a)-(d).

[55] 33 U.S.C. § 2713(e); 33 CFR § 136.229.

[56] 33 U.S.C. § 2702(b)(2)(E).

[57] 33 U.S.C. § 2701(20).

[58] 33 U.S.C. § 2713(e); 33 CFR §§ 136.233(a)-(d).
[59] 33 U.S.C. § 2713(e); 33 CFR § 136.235.
[60] 33 U.S.C. § 2702(b)(2)(F).
[61] 33 U.S.C. § 2702(b)(2)(F).
[62] 33 U.S.C. § 2713(e); 33 CFR §§ 136.239(a)-(d).
[63] 33 U.S.C. § 2713(e); 33 CFR § 136.241.
[64] 33 U.S.C. § 2703(a). For the purposes of determining whether the "act of God" complete defense is available to a responsible party, the term "act of God" means an "unanticipated grave natural disaster or other natural phenomenon of an exceptional, inevitable, and irresistible character the effects of which could not have been prevented or avoided by the exercise of due care or foresight." 33 U.S.C. § 2701(1). In the event that the upcoming hurricane season exacerbates the Deepwater Horizon oil spill, the question arises as to whether a hurricane in the Gulf of Mexico could be considered an "unanticipated" natural phenomenon. Analysis and consideration of this question, however, are beyond the scope of this CRS report.
[65] 33 U.S.C. §§ 2703(a)(3)(A) and (B).
[66] 33 U.S.C. § 2703(c).
[67] 33 U.S.C. § 2703(d)(5).
[68] 33 U.S.C. § 2712. The standards and procedural requirements for claims filed against the Fund are set forth in the Coast Guard's OPA regulations. *See* 33 C.F.R. §§ 136.1 through 136.241.
[69] 26 U.S.C. §§ 4611(a)(1) and (2). The Oil Spill Liability Trust fund is also financed by a per-barrel tax on domestic crude oil "used in or exported from the United States." 26 U.S.C. § 4611(b)(1)(A).
[70] 26 U.S.C. § 9509(c)(2)(A).
[71] 26 U.S.C. § 9509(d)(3)(B). Senate bill S. 3306 would reinstate the Fund's authority to receive advances from the Treasury, if enacted. The House companion bill to S. 3306, H.R. 5214, would amend the Fund's authority to borrow in the same manner as S. 3306, if enacted.
[72] 33 U.S.C. § 2715(c).

In: Gulf Oil Spill of 2010...
Editors: C. R. Walsh, J. P. Duncan

ISBN: 978-1-61324-729-7
© 2012 Nova Science Publishers, Inc.

Chapter 3

LIABILITY AND COMPENSATION REQUIREMENTS UNDER THE OIL POLLUTION ACT[*]

National Commission on the BP Deepwater Horizon Oil Spill and Offshore Drilling

STAFF WORKING PAPER NO. 10.

Staff working papers are written by the staff of the National Commission on the BP Deepwater Horizon Oil Spill and Offshore Drilling for the use of members of the Commission. They do not necessarily reflect the views of the Commission or of any of its members. In addition, they may be based in part on confidential interviews with government and non-government personnel.

I. ISSUE

In November 2010, BP estimated that its total costs from the *Deepwater Horizon* spill, including the clean-up, penalties and damages, will total nearly

---

[*] This is an edited, reformatted and augmented version of a National Commission on the BP Deepwater Horizon Oil Spill and Offshore Drilling publication, Staff Working Paper No. 10.

forty billion dollars.[1] BP has the resources to pay this enormous sum. Those who have suffered individual damages from the spill and those who wish to see the Gulf's natural resources restored are fortunate that BP, rather than a smaller oil and gas company, was responsible for the spill. However, the fact that BP is able to provide full monetary compensation for damages that it causes is no more than a fortuity, not a product of regulatory design. If a company with less financial means had caused the spill,[2] the company would likely have declared bankruptcy long before paying anything close to the damages caused.[3]

As discussed below, the current law limits the amount of liability for damages caused by oil spills. As a result, it provides little incentive for improving safety practices to decrease the likelihood of major spills, and it limits the ability of those of who suffer damages to receive full compensation. In the immediate aftermath of the BP spill, several legislative proposals were introduced to change the applicable liability caps and the financial responsibility provisions of the Oil Pollution Act of 1990, as a way to address both of these problems.

The purpose of this paper is to provide background information to the Commission in support of the Commission's consideration of policy options related to liability caps and financial responsibility applicable to oil spills from offshore facilities. To that end, the paper briefly summarizes existing law and identifies some of the more significant policy issues raised concerning possible amendment of current law. The paper discusses each of these issues, highlighting some of the competing concerns implicated by different policy outcomes.[4]

II. EXISTING LAW

A. The Oil Pollution Act of 1990

Under the Oil Pollution Act of 1990 "responsible parties," including lessees of offshore facilities,[5] are strictly liable for removal costs and certain damages resulting from a spill, subject to caps on liability.[6] Responsible parties are not liable for the costs of removal or damages if violations are caused solely by an act of God, act of war, or act or omission of a third party.

Lessees are required to demonstrate financial responsibility in an amount between $35 million and $150 million. [7] The Bureau of Ocean Energy Management, Regulation and Enforcement ("BOEMRE," formerly the

Minerals Management Service) defines "Oil Spill Financial Responsibility" as "the capability and means by which a responsible party for a covered offshore facility will meet removal costs and damages for which it is liable" under the Oil Pollution Act.[8] BOEMRE's regulations establish guidelines for the level of financial responsibility necessary, based on the estimated worst-case discharge from offshore facilities. "Worst case discharge" for a well is defined as four times the estimated uncontrolled flow volume for the first 24 hours of a spill, as set forth in the responsible party's response plan.[9] BOEMRE has authority to increase the required amount based on relevant operational, environmental, human health or other risks posed by the operation,[10] but as discussed below, the total amount required to be demonstrated may not exceed $150 million.[11] Firms may demonstrate financial responsibility in various ways, including surety bonds, guarantees, letters of credit, and self-insurance; the most common method is through an insurance certificate.[12]

Finally, certain claims for natural resource damages and "uncompensated damages" can be made to, and paid out of, the Oil Spill Liability Trust Fund (Trust Fund or Fund). The Trust Fund is currently funded by an 8 cent per barrel tax on domestic production and imported oil.[13]

B. Current Limitations on Liability and Compensation

The Oil Pollution Act currently limits liability and compensation for damages caused by a spill from on offshore facility in three ways. First, it caps liability for damages from a spill from an offshore facility at $75 million per incident.[14] This limit does not apply if the incident was proximately caused by a responsible party's gross negligence, willful misconduct, or violation of applicable Federal safety, construction, or operation regulation.[15] The limitation on liability also does not apply to civil and criminal penalties under federal and state law, oil spill removal costs under federal law, or claims for damages brought under state law.[16]

Second, under the Oil Pollution Act, the highest level of financial responsibility a covered facility must demonstrate is $150 million, the amount required for facilities whose worst-case discharge volume exceeds 105,000 barrels.[17] Thus, even though an offshore facility is potentially liable for damages that exceed $75 million (for example, in the event that the responsible party acted with gross negligence), it is not required to demonstrate actual capacity to pay damages beyond $150 million.

Third, if the responsible party is not able to compensate all of the damages caused by the spill, the Trust Fund is available to cover certain damages.[18] However, the amount authorized per incident is limited to $1 billion[19] and, until recently, the overall limit on the Fund was $2.7 billion.[20] As of June 2010, the Fund's balance was approximately $1.5 billion.[21]

Thus, in the case of a large spill, there is no certainty under current law that a company would have the financial means to fully compensate victims of the spill. Moreover, the Trust Fund would likely not provide sufficient backup, and a significant portion of the injuries caused to individuals and natural resources as well as government response costs could go uncompensated.

III. AMENDING EXISTING LAW

A. Legislative Proposals

During the 111th Congress, members introduced bills that would have addressed, in different ways, the unfavorable impacts of limitations on the liability cap and financial responsibility requirements. The bills contain provisions that would do some or all of the following:

- Eliminate the liability cap for offshore facilities[22]
- Change the financial responsibility requirements by raising limits or requiring the Secretary of the Interior to review requirements[23]
- Require participation in a mutual liability pool[24]
- Increase the amount of available per incident funding in the Trust Fund[25]

Congress, however, was unable during the 111th Congress to reach the compromise necessary to secure passage of any relevant legislation.

B. Relevant Considerations in Developing Policy Options

Raising or eliminating the liability cap and increasing financial responsibility serve two distinct, important policy goals. First, changing existing law in this manner could create stronger incentives for firms to internalize risk and operate more safely offshore. And second, such steps would provide greater assurances that, in the event of a major spill, there would be adequate funds available to compensate for damages and costs caused by such spills, without requiring the taxpayer to bear the burden of compensation.

There are a variety of ways that existing law could be modified to further these overall objectives:

- Raising the liability cap, using a phased in approach
- Raising financial responsibility requirements, using a phased in approach
- Ensuring an evaluation of risk by the regulator in setting criteria for financial responsibility levels, and/or by insurance companies in determining premiums
- Increasing the per-incident limits on payout from the Oil Spill Liability Trust Fund

Relationship of liability caps and financial responsibility limits

At the outset, it is important to note that it is unlikely that raising or eliminating the liability cap will have the desired effect of providing incentives for safe practices or ensuring full compensation for victims, unless demonstrated financial responsibility is required at levels commensurate with the cap.[26] The debate over the Oil Pollution Act liability cap has focused primarily on increasing or eliminating the liability cap itself; discussion of financial responsibility requirements has been secondary. However, if the liability cap is increased without a corresponding increase in financial responsibility requirements, then a firm could meet its financial responsibility requirements and still go bankrupt before paying even a small fraction of the damage associated with a spill. The liability limit would, in effect, be irrelevant.[27] Some have argued that the financial responsibility requirement should be *higher* than the relevant liability cap, in order to ensure that the responsible party is capable of paying the full range of damages, costs and penalties applicable under federal and state law.[28]

1. General Policy Reasons for Raising the Cap and Increasing Financial Responsibility Requirements for Offshore Facilities

Incentives

One argument frequently advanced for raising liability caps is that significant potential monetary liability increases a company's incentive to improve its safety practices.[29] To the extent that a liability scheme provides incentives to internalize costs, the comparatively low $75 million cap distorts companies' incentives to engage in practices that prevent spills.[30] This point has been made by numerous economists who have reviewed the Oil Pollution Act liability cap. Under basic economic theory, companies that have the potential to cause significant harm should pay for the costs they inflict; economists have thus opined that the best way to ensure internalization of risk is to require strict, unlimited liability for all damages inflicted on the public by an accident.[31] For example, in testimony before Congress, MIT economist Michael Greenstone stated,

> [T]he $75 million cap on liabilities for economic damages means that oil companies do not bear full responsibility for oil spills. This misalignment of incentives is a classic case of moral hazard. Firms and people behave differently when they are protected from the consequences of their actions... Market forces require [oil companies] to make decisions about where to drill and which safety equipment to use, based on benefit-cost analyses of the impact on their bottom line. . . If the expected cost benefits exceed the expected costs, the decision to move forward will appear sound.[32]

Therefore, according to Professor Greenstone, the offshore drilling liability cap "inevitably distorts" the way companies make decisions to drill.[33]

The incentive argument is somewhat diminished, however, by the fact that there are significant limitations on the scope of the liability cap's applicability. As noted above, caps do not apply to removal costs, damage claims under state law, and penalty actions, and they do not apply where there has been gross negligence, willful misconduct, or violation of applicable Federal safety, construction, or operation regulation. Thus, in the case of the *Deepwater Horizon* spill, for example, BP and/or other responsible parties may be liable for removal costs, the potential billions of dollars in civil and criminal penalties, unlimited liability for damages in some states and potentially, civil and/or criminal penalties under state law.[34] In any event, to the extent that a liability cap is not waived, the aggregate expected damages from a spill are lower than they otherwise would be, and this fact may have an effect on a company's incentive to adopt more stringent safety practices.[35]

Compensation

In addition, increased liability and financial responsibility will help insure that individuals, property owners and natural resources trustees who suffer damages but have no role in causing a spill are fully compensated. As of

December 2, 2010, BP had already pledged or paid billions of dollars in damages, including through the establishment of a $20 billion fund for private claimants and through natural resource damage assessment payments to states.[36] As noted above, if BP had not agreed to waive the liability cap, and the cap was not subject to one of the exceptions enumerated in Oil Pollution Act, then innocent victims would have been out of luck, or the taxpayer would have borne the burden of compensating those victims.

2. Potential Impact on the Insurance Industry

In testimony before Congress, insurance officials and others have warned of unintended consequences of an increase in, or elimination of, the liability cap. With respect to the short term, industry sources have testified that the offshore energy insurance market currently has a finite amount of liability insurance capacity -- in the range of $1 billion to $1.5 billion per company.[37] One insurance industry representative described a range of obstacles for insurers and purchasers of insurance, including but not limited to: 1) that the entire global energy market is limited to $3 billion in premiums; 2) that higher liability limits would increase the demand for coverage, exhausting available capacity; and 3) that underwriting for low probability, high severity events is challenging for insurers and reinsurers.[38] Similarly, the Congressional Research Service, based upon a review and study of the offshore drilling insurance business, outlined potential consequences for the insurance market that would result from increase or removal of liability limits. These include: 1) the emergence of a "hard" market –high prices and limited coverage – in contrast to the "soft" insurance market for offshore liability prior to the spill;[39] 2) higher costs of insurance based upon higher strict liability limits, resulting in some operators choosing to selfinsure;[40] and 3) reluctance of private insurers to commit capital to undefined risks, based on unknown legislative changes and litigation risk.[41]

It is also possible that the market will eventually adjust to the new liability regimes. According to the extensive Congressional Research Service analysis, the insurance market would likely support the use of alternative sources of insurance capacity, such as "reinsurance sidecars," catastrophe bonds or energy insurance financial futures and options, to spread risk and increase available capital in the insurance market.[42] Another commentator has suggested that insurance capacity may also increase if there is a shift towards writing insurance policies for facilities, rather than for companies.[43] In short, the potential impact on the insurance industry is uncertain; for this reason, an approach that phases in changes to the liability cap and financial responsibility requirements may be desirable.

3. Potential Impact on the Structure of the Offshore Drilling Industry

Representatives of the offshore drilling and insurance industries have predicted that, in light of the current lack of availability of insurance described above, an increase in the liability cap, and certainly elimination of the liability cap, will result in an exodus of smaller, independent companies from offshore drilling operations because they would not be able to obtain the insurance necessary to operate.[44] They advance several arguments for ensuring that legislation does not lead to such a result.

First, the independents develop many smaller and end of life fields that the larger firms find uneconomic or inefficient[45]; thus, excluding the independents would probably result in reduced production. Second, the exit of some businesses would decrease the competition for lease sales, most likely resulting in a decrease in the amount of money the government receives for lease sales.[46] Third, excluding the independents from drilling and production would have significant negative impacts on employment and economic activity in the Gulf oil states.[47] Finally, if the elimination of or increase in the cap were applied retroactively it could cause operators to relinquish leases, which would in turn result in a decline in production in some areas.[48] These points underscored the sensitivity of the liability cap issue in early congressional debates; and were at least one reason that the Senate rejected the Reid Clean Energy legislation.

There are also counterarguments to the proposition that policy reform should be aimed at ensuring the ability of small companies to remain in the offshore drilling business. Economists have argued that the market should drive which businesses engage in the drilling business, and if small companies cannot afford to drill safely, they should not do so.[49] There is also some evidence that the effect on employment would not actually be substantial because there would simply be a shift in the companies that engage in drilling operations and provide jobs.[50]

4. Possible Ways to Mitigate Adverse Impact on Smaller, Independent Companies of Raising the Liability Cap and Increasing Financial Responsibility Requirements

The impact on smaller, independent companies could potentially be mitigated in several ways. Options include:

- Raising but not eliminating the liability cap.
- Requiring pooling of risk in a mutual fund, as proposed in the Landrieu bill, which requires participation in a mutual fund. One downside to the mutual liability pool is that it decreases incentives for individual firms to improve safety practices. This problem could potentially be addressed by tying premium levels to financial and safety risk posed by an individual company's activities. The Landrieu bill directs the Secretary of the Interior to determine how to set premiums.[51]
- Phasing in of financial responsibility requirements until the insurance industry adjusts to the demand for insurance and/or new financial products are created.[52]

Adverse effects on small companies could also be offset by partnering with firms with deeper pockets. "Joint ventures" between larger and smaller companies already exist, and a policy change is probably not necessary to encourage such arrangements.

5. Risk Evaluation: How and Who

Taking Risk into Account

BOEMRE currently determines level of financial responsibility based upon potential worst case discharge, as required by the Oil Pollution Act. Although this analysis to some degree accounts for risk associated with individual drilling activities, it does not fully account for the range of factors that could affect the cost of a spill. Accordingly, in the regulatory context, staff has advised that the Commission may wish to recommend that BOEMRE require more stringent management of geological, environmental and operational risk, to ensure that the regulatory scheme better responds to potential consequences of spills, particularly in high-risk and frontier areas. Similarly, the Commission may decide to recommend that BOEMRE consider specific criteria relevant to a determination of risk, when establishing financial responsibility limits applicable to a particular company or facility. Risk criteria could include, at the least: geological and environmental considerations, the applicant's experience and expertise, and applicable risk management plans. This increased scrutiny would provide an additional safeguard against unqualified companies entering the offshore drilling market.[53]

Congress has recently considered a variety of related reforms. The CLEAR Act, for example, raised the Oil Pollution Act's maximum financial responsibility to $300 million. But it allowed the regulator to prescribe a lower amount, for a responsible party, based on the following criteria: (i) the market capacity of the insurance industry to issue such instruments; (ii) the operational risk of a discharge and the effects of that discharge on the environment and the region; (iii) the quantity and location of the oil and gas that is explored for, drilled for, produced, or transported by the responsible party; (iv) the asset value of the owner of the offshore facility, including the combined asset value of all partners that own the facility; (v) the cost of all removal costs and damages for which the owner may be liable under this Act based on a worstcase-scenario; (vi) the safety history of the owner of the offshore facility; (vii) any other factors that the President considers appropriate. [54]

Similarly, a bill introduced in the Senate would require the regulator to determine liability limits for offshore facilities, based upon: (i) the water depth of the lease; (ii) the minimum projected well depth of the lease; (iii) the proximity of the lease to oil and gas emergency response equipment and infrastructure; (iv) the likelihood of the offshore facility covered by the lease to encounter broken sea ice; (v) the record and historical number of regulatory violations of the leaseholder under the Outer Continental Shelf Lands Act (43 U.S.C. 1331 et seq.) or the Federal Water Pollution Control Act (33 U.S.C. 1251 et seq.) (or the absence of such a record or violations);(vi) the estimated hydrocarbon reserves of the lease; (vii) the estimated well pressure, expressed in pounds per square inch, of the reservoir associated with the lease; (viii) the availability and projected availability of funds in the Oil Spill Liability Trust Fund established by section 9509 of the Internal Revenue Code of 1986;(ix) other available remedies under law;(x) the estimated economic value of nonenergy coastal resources that may be impacted by a spill of national significance involving the offshore facility covered by the lease;(xi) whether the offshore facility covered by the lease employs a subsea or surface blowout preventer stack; and (xii) the availability of industry payments.[55]

Roles of Government and Private Entities

There are clear benefits to having an external party evaluate risk and monitor individual firms' safety compliance. As noted above, it is appropriate for the government (BOEMRE) to play this role, to promote enhanced risk management in offshore operations and to discourage unqualified companies from performing offshore drilling operations. There is also a role for private insurance companies in evaluating risk. Unlike the government, insurance companies have both the financial incentives and the resources to monitor risk-taking and lower their own exposure.[56] If liabilities are borne by insurance carriers, carriers will also have a strong incentive to promote new safety techniques and methods by encouraging other institutions (including insured firms) to engage in such research. They may also require certification from private firms specializing in risk management.[57]

6. Providing for Full Victim Compensation

If liability and financial responsibility limits are not set at a level that will ensure payment of damages for all spills, then another source of funding will be required to fully compensate victims of a spill. The federal government could pay additional compensation costs, but this approach requires the taxpayer to foot the bill, essentially subsidizing the drilling activity.

If the Oil Spill Liability Trust Fund per-incident limit is raised, then the costs are essentially borne by those who benefit from oil production activities. Reliance on the Trust Fund to fully compensate victims would not necessarily provide an incentive to offshore facilities to mitigate risks because risks are pooled. Such a reduction in incentives counsels against having the Trust Fund be the sole source of compensation for damages in the event of a big spill. However, raising the per-incident payout limit would help ensure that victims have access to compensation without the need to seek further specific funding from Congress or otherwise burden the taxpayer.

Note that currently, there is no overall cap on the Trust Fund. From 2005 – 2008, there was a $2.7 billion cap on the Fund, but it was removed as part of the Stimulus package in 2008. If the per-incident limits are raised, it will be important 1) that the Fund be allowed to continue to grow, and 2) that if a cap is again imposed on the Fund, the cap is significantly higher than the established per-incident limit on payouts.

End Notes

[1] Graeme Wearden, "BP oil spill costs to hit $40bn," *The Guardian*, November 2, 2010.

[2] In the Gulf, the oil exploration and production industry is composed of two main groups of operators: (a) international and major integrated companies (the "majors"), and (b) independent exploration and production companies (the "independents"), who themselves are often large companies. *See* Rawle King, *Deepwater Horizon Oil Spill Disaster: Risk, Recovery, and Insurance Implications*, (Washington, D.C.: Congressional Research Service, July 12, 2010), 5; IHS Global Insight (USA), Inc., *The Economic Impact of the Gulf of Mexico Offshore Oil and Natural Gas Industry and the Role of Independents* (July 21, 2010), 4.

[3] *See, e.g.,* Mark Cohen et al., *Deepwater Drilling: Law, Policy and Economics of Firm Organization and Safety,* (Resources for the Future, January 2011), 33.

[4] For further discussion of the issues addressed herein, please refer to: Cohen et al., *Deepwater Drilling: Law, Policy and Economics of Firm Organization and Safety* and Thomas Merrill, *Insurance and Safety Incentives* (2010). These papers are posted on the Commission's website, www.oilspillcommission.gov.

[5] This discussion is limited to the liability cap for offshore facilities, although OPA covers liability for vessels and other sources as well.

[6] 33 U.S.C. § 2702. Compensable damages are damages for: natural resources; real or personal property; subsistence use; revenues; profits and earning capacity; and public services.

[7] 33 U.S.C. § 2716(c).

[8] 30 C.F.R. § 253.3.

[9] 30 C.F.R. § 253.14(a)(1).

[10] 30 C.F.R. § 253.13.

[11] *Id.*

[12] 30 C.F.R. § 253.20-32; King, *Deepwater Horizon Oil Spill Disaster: Risk, Recovery, and Insurance Implications,* 7. For a more detailed discussion of how firms currently demonstrate financial responsibility *see id.,* Merrill, *Insurance and Safety Incentives,* 7-9.

[13] 33 U.S.C. § 2712 (a)(4). In October 2008, Congress raised the tax per barrel from $0.05 to $0.08 until January 1, 2017, and to $0.09 from January 1, 2017 to December 31, 2017. Congress also removed the $2.7 billion cap on the Fund. Emergency Economic Stabilization Act of 2008, P.L. 110-343, 122 Stat. 3860; *compare* 26 U.S.C. § 4611 (2007) *with* 26 U.S.C. § 4611 (2010).

[14] 33 U.S.C. § 2714(a)(3).

[15] *Id.* § 2704(c)(1).

[16] *Id.* § 2718. See Cohen et al, *Deepwater Drilling: Law, Policy and Economics of Firm Organization and Safety,* 30-31.

[17] 33 C.F.R § 253.14.

[18] 33 U.S.C. § 2712.

[19] 26 U.S.C. § 9509(c)(2).

[20] Even where the Fund gets reimbursement from the responsible party, the amounts the Fund initially paid out are counted against the $1 billion limit. Thus, in the BP/*Deepwater Horizon* case, even though BP has reimbursed the Fund for over $500 million spent on government response activities, the fact that the Fund was used for those response activities means that less than $500 million is available to compensate damages going forward. Government Accountability Office, *GAO-11-90R Deepwater Horizon Oil Spill: Preliminary Assessment of Federal Financial Risks and Cost Reimbursement and Notification Policies and Procedures* (November 12, 2010). This issue could be addressed in legislation.

[21] Hearing on Liability and Financial Responsibility for Oil Spills under the Oil Pollution Act of 1990 and Related Statutes, Before H. Comm. on Transportation and Infrastructure, 111th Cong. (2010) (statement of Craig Bennett, Director, National Pollution Funds Center), 3.

[22] Consolidated Land, Energy, and Aquatic Resources Act, H.R. 3534, 111th Cong. ("CLEAR Act"). § 702 (2010) (as passed by House); S. 3663, 111th Cong. §102 (2010)("Reid Clean Energy bill"); Restoring Ecosystem Sustainability and Protection on the Delta Act, S. 3763, 111th Cong. (2010) S. 3763, 111th Cong. § 6 (2010) ("RESPOND Act" or "Landrieu bill") (liability between $250 million and $10 billion is borne by mutual liability pool; liability over $10 billion is borne by responsible party).

[23] CLEAR Act, § 703 (raises financial responsibility for offshore facilities to $300 million; may be less based on certain criteria); Reid Clean Energy bill, S. 3663, § 306 (every five years, the Secretary of the Interior will review minimum financial responsibility requirements, make adjustments for inflation, and make recommendations to Congress on financial responsibility requirements.)

[24] RESPOND Act, § 7 (pool provides insurance for costs between $250 million and $10 billion; premiums are based on amounts set by the Secretary of the Interior).

[25] Reid Clean Energy bill, S. 3663, § 5001 (Increases amount available per incident to $5 billion; increases per barrel tax to 45 cents). *See also* Hearing on Liability and Financial Responsibility for Oil Spills under the Oil Pollution Act of 1990 and Related Statutes, Before the Comm. on Transportation and Infrastructure, 111th Cong. (2010) (statement of Thomas Perrelli, Associate Attorney General) (describing Obama Administration proposal that increases per incident limit).

[26] *See* Cohen et al., *Deepwater Drilling: Law, Policy and Economics of Firm Organization and Safety*, 45.

[27] *Id.*

[28] *Id.*

[29] *Id.*, 40; Merrill, *Insurance and Safety Incentives*, 6.

[30] Merrill, *Insurance and Safety Incentives*, 6.

[31] *See* Ishan Nath, *Economists' Perspectives on Liability Caps and Insurance for the Offshore Oil and Gas Industry in the Wake of the Macondo Blowout* (2010), 5-8, posted on www.oilspillcommission.gov(quoting numerous academic economists who stated that the best way to ensure internalization and mitigation of risk is through removal of Oil Pollution Act liability caps).

[32] Hearing on Liability and Financial Responsibility for Oil Spills under the Oil Pollution Act of 1990 and Related Statutes, Before House Comm. on Transportation and Infrastructure, 111th Cong. (2010) (Statement of Michael Greenstone, MIT), 3.

[33] *Id.*

[34] *See* Cohen et al., *Deepwater Drilling: Law, Policy and Economics of Firm Organization and Safety*, 29-31. *See* The Big Oil Bailout Prevention Liability Act of 2010: Hearing on S. 3305 before the S. Comm. on Environment and Public Works, 111th Cong. (2010) (statement of Ken Murchison, Louisiana State University), 7-8.

[35] Cohen et al., *Deepwater Drilling: Law, Policy and Economics of Firm Organization and Safety*, 31. *See also* Greenstone, 4 ("We cannot know whether the result would have been different without the cap, but what is clear is that there were economic incentives for companies to cut corners. Those incentives will remain as long as the cap is set at such a low level relative to the risk").

[36] BP, Claims and Government Payments Gulf of Mexico Oil Spill Public Report (December 2, 2010).

[37] *See, e.g.,* Hearing on Liability and Financial Responsibility for Oil Spills under the Oil Pollution Act of 1990 and Related Statutes, Before the H. Comm. on Transportation and Infrastructure, 111th Cong. (2010)(statement of Dr. Robert Hartwig, President and Economist, Insurance Information Institute), 13; *see* King, *Deepwater Horizon Oil Spill Disaster: Risk, Recovery, and Insurance Implications*, 2.

[38] Hearing on Liability and Financial Responsibility for Oil Spills under the Oil Pollution Act of 1990 and Related Statutes, Before the H. Comm. on Transportation and Infrastructure, 111th Cong. (2010)(statement of Dr. Robert Hartwig, President and Economist, Insurance Information Institute), 13.

[39] King, *Deepwater Horizon Oil Spill Disaster: Risk, Recovery, and Insurance Implications*, 17.

[40] *Id.*, 16 ("Operators may find themselves assuming or retaining higher levels of self-insurance, which might affect the BOEMRE's offshore oil and gas leasing bidding and ultimately the royalties earned for the U.S. Treasury.")

[41] *Id.*, 15-18.

[42] *Id.*, 17-18.

[43] Merrill, *Insurance and Safety Incentives*, 12 (suggesting that if insurance companies focus more on risk, they may offer policies designed for specific facilities; in turn, greater diversification of risk would induce greater insurance capacity).

[44] Hearing on Liability and Financial Responsibility for Oil Spills under the Oil Pollution Act of 1990 and Related Statutes, Before the H. Comm. on Transportation and Infrastructure, 111th Cong. (2010)(statement of Charles Anderson, SKULD North America); Liability and Financial Responsibility for Oil Spills under the Oil Pollution Act of 1990 and Related Statutes: Hearing Before the H. Comm. on Transp. & Infrastructure, 111th Cong. (June 9, 2010) (Statement of Jack Gerard, President and CEO, American Petroleum Institute).

[45] Scott Gutterman (President and CEO, LLOG Exploration Co, LLC), interview with Commission staff, November 19, 2010.

[46] Liability and Financial Responsibility for Oil Spills under the Oil Pollution Act of 1990 and Related Statutes: Hearing Before the H. Comm. on Transp. & Infrastructure, 111th Cong. (June 9, 2010) (Statement of the Honorable James L. Oberstar), 3; Cohen et al, *Deepwater Drilling: Law, Policy and Economics of Firm Organization and Safety*, 42.

[47] IHS Global Insight (USA), Inc., *The Economic Impact of the Gulf of Mexico Offshore Oil and Natural Gas Industry and the Role of Independents*, 5-6.

[48] *See* King, *Deepwater Horizon Oil Spill Disaster: Risk, Recovery, and Insurance Implications*, 19.

[49] Hearing on Liability and Financial Responsibility for Oil Spills under the Oil Pollution Act of 1990 and Related Statutes, Before House Comm. on Transportation and Infrastructure, 111th Cong. (2010) (Statement of Michael Greenstone, MIT), 5; Nath, *Economists' Perspectives on Liability Caps and Insurance for the Offshore Oil and Gas Industry in the Wake of the Macondo Blowout*, 18 (quoting economists Kenneth Arrow and Don Fullerton)

[50] Nath, *Economists' Perspectives on Liability Caps and Insurance for the Offshore Oil and Gas Industry*, 19-21 (quoting numerous economists who reject the theory that the impact of unlimited liability on small firms would lead to job loss).

[51] The Landrieu proposal applies to all companies; one potential modification, which would be favored by the majors, would be to allow the largest companies to opt out and self insure.

[52] *See* Merrill, *Insurance and Safety Incentives*, 6.

[53] Greater regulatory focus on risk management would also likely result in increased insurance capacity, since presumably insurance companies will be more willing to provide insurance where risks are lower. Merrill, *Insurance and Safety Incentives*, 15.

[54] CLEAR Act, § 703.

[55] Oil Spill Compensation Act of 2010, S. 3542, 111th Cong. (2010), §301(e).

[56] *See* Cohen et al, *Deepwater Drilling: Law, Policy and Economics of Firm Organization and Safety*, 36-37.

[57] *See* Merrill, *Insurance and Safety Incentives*, 15-16.

In: Gulf Oil Spill of 2010...
Editors: C. R. Walsh, J. P. Duncan

ISBN: 978-1-61324-729-7
© 2012 Nova Science Publishers, Inc.

Chapter 4

THE 2010 OIL SPILL: NATURAL RESOURCE DAMAGE ASSESSMENT UNDER THE OIL POLLUTION ACT[*]

Kristina Alexander

SUMMARY

The 2010 *Deepwater Horizon* oil spill leaked an estimated 4.1 million barrels of oil into the Gulf of Mexico, damaging the waters, shores, and marshes, and the fish and wildlife that live there. The Oil Pollution Act (OPA) establishes a process for assessing the damages to those natural resources and assigning responsibility for restoration to the parties responsible. BP was named the responsible party for the spill. The Natural Resources Damage Assessment (NRDA) process allows Trustees of affected states and the federal government (and Indian tribes and foreign governments, if applicable) to determine the levels of harm and the appropriate remedies.

The types of damages that are recoverable include the cost of replacing or restoring the lost resource, the lost value of those resources if or until they are recovered, and any costs incurred in assessing the harm. Claims by individuals or businesses are not allowed, as all injuries are to the resources managed by state, federal, tribal, or foreign governments.

---

[*] This is an edited, reformatted and augmented version of a Congressional Research Service publication, CRS Report for Congress R41396, from www.crs.gov, dated September 8, 2010.

OPA allows recovery from the responsible parties for harm resulting from response efforts, which in this case could include in situ burning, use of dispersants, and vehicle traffic on shores and marshes. The $20 billion escrow fund set up by BP in June 2010 is not for government NRDA claims, but it can be used to reimburse individual losses of subsistence use of natural resources, primarily lost fishing opportunities, which are covered by OPA.

Under NRDA, Trustees design a recovery plan that is paid for or implemented by any responsible parties. If the responsible parties refuse to pay or cannot reach an agreement with the Trustees, the Trustees can sue the responsible party under NRDA for those damages or seek compensation from the Oil Spill Liability Trust Fund, but there is a cap of $500 million for natural resources damage. The federal government can then seek restitution from the responsible parties for the sums taken from that fund. OPA caps liability for offshore drilling units at $75 million for economic damages, but does not limit liability for removal costs.

Both the caps on the Oil Spill Trust Fund and on OPA have captured Congress's attention, as has Gulf restoration. H.R. 3534 would remove the OPA cap on damages for offshore facilities. It would also establish a task force to create a restoration plan within 12 months of enactment. This plan appears to be separate from the restoration plan under NRDA. However, Title V of H.R. 3534 overlaps parts of the NRDA process.

INTRODUCTION

Natural resources are protected by the government under a long-standing common law tradition known as the public trust doctrine, which dates back centuries before the United States was created. Under the public trust doctrine, natural resources belonging to the government are to be managed for the benefit of all. Within the United States, this means that, for the most part, management of the natural resources in the public trust falls to the states, except where a statute puts the federal government in control. For example, while wildlife management is a state responsibility, the Endangered Species Act, the Marine Mammal Protection Act, and the Migratory Bird Treaty Act all bring certain species under federal protection.

When resources in the public trust are harmed by contamination, federal, state, foreign, and tribal governments may seek compensation for damage to natural resources under certain laws. This is done in two steps: first, by assessing the harm; then, by determining how and what restoration will take place. Compensation for natural resource damage is intended to restore the natural resources to their condition before the damage and to compensate the public for the lost use of those resources. The estimated 4.1 million barrels[1] of oil released during the 2010 *Deepwater Horizon* oil spill have had and will continue to have an impact on the natural resources of the Gulf region.

Statutory Authority

Natural Resource Damage Assessment and Recovery (NRDA) is authorized by several statutes, depending on the type of contamination: the Comprehensive Environmental Response, Compensation, and Liability Act (CERCLA);[2] the Clean Water Act (CWA);[3] the Oil Pollution Act of 1990 (OPA);[4] the National Marine Sanctuaries Act;[5] and the Park System Resources Protection Act.[6] Each statute allows collecting money as compensation for natural resource damages. Any recovery under these schemes must go toward restoration of injured resources. NRDA does not directly assist individuals affected by an oil spill and does not provide for punitive damages. The NRDA process for the 2010 oil spill in the Gulf of Mexico will be conducted pursuant to OPA. Accordingly, this report will only address that statute.

OPA (sometimes known as OPA 90) applies to discharges of oil into or on the navigable waters of the United States, adjoining shorelines, and the exclusive economic zone of the United States (where the BP spill was located).[7] It was enacted due in part to the *Exxon Valdez* spill in 1989, where liability was imposed primarily via the CWA. OPA amended the CWA and several other statutes imposing oil spill liability to create a unified oil spill liability regime, expand the coverage of such statutes, increase liability, strengthen federal response authority, and establish a fund to ensure that claims are paid up to a stated amount. It has been held to preempt other maritime remedies.[8] As with the CWA, liability under OPA is strict, and joint and several.[9] Under OPA, each *responsible party* for an oil spill is liable for removal costs and six specified categories of economic damages.[10] One of these categories is natural resource damages, replacing the preexisting natural resource damages provisions in the CWA for oil spills.[11] OPA defines natural resource damages as "[d]amages for injury to, destruction of, loss of, or loss of use of, natural resources, including the reasonable costs of assessing the damage, which shall be recoverable by a United States trustee, a State trustee, an Indian tribe trustee, or a foreign trustee."[12]

Natural resources are defined broadly by the act to include the following: "land, fish, wildlife, biota, air, water, ground water, drinking water supplies, and other such resources belonging to, managed by, held in trust by, appertaining to, or otherwise controlled by the United States (including the resources of the exclusive economic zone), any State or local government or Indian tribe, or any foreign government."[13]

The National Oceanic and Atmospheric Administration (NOAA) of the Department of Commerce oversees the NRDA process under OPA. Its regulations are found at 15 C.F.R. part 990. It also may act as a Trustee when the resources it protects are harmed, such as in this case. Currently, NOAA is involved in 13 other NRDA oil spill cases in the Gulf besides the BP spill.[14]

Trustees

The governments in charge of the resources—federal, state, tribal, and foreign—are known as Trustees under NRDA. They coordinate the process of determining the extent of damages, the value of the resources, and the method(s) of restoration, including compensation amounts. They are charged with acting "on behalf of the public."[15] By establishing a collaborative process for resolving liability issues, NRDA is designed to avoid litigation. According to discussion in the *Congressional Record* about OPA, "[OPA] is intended to allow for quick and complete payment of reasonable claims without resort to cumbersome litigation."[16] Litigation may be avoided altogether if the responsible parties consent to the Trustees' plan for assessment and restoration.

Additionally, OPA requires presenting NRDA claims to the responsible parties before any suit can be filed or other action taken.[17] This allows a chance for a pre-court settlement.

For the 2010 oil spill, the federal government Trustees include the Fish and Wildlife Service and National Park Service of the Department of the Interior and NOAA. The Federal Lead Administrative Trustee is the Department of the Interior. The state Trustees are the governors and various agencies of the states affected by the spill: Alabama, Florida, Louisiana, Mississippi, and Texas.[18] Indian tribes may be Trustees for affected tribal lands, but no such property has been identified as injured from the spill. No foreign governments have been affected yet, but it remains a possibility: Canada might have a claim if the habits of migratory birds are disrupted; damage to Mexican resources is a possibility.

Table 1. Trustees in Gulf NRDA Process (as of August 2010)

Department of the Interior	Department of Commerce	State of Louisiana	State of Mississippi	State of Alabama	State of Florida	State of Texas
U.S. Fish and Wildlife Service	National Oceanic and Atmospheric Administration	Coastal Protection and Restoration Authority	Department of Environmental Quality	Department of Conservation and Natural Resources,	Department of Environmental Protection	Parks and Wildlife Department
National Park Service		Oil Spill Coordinator's Office		Geological Survey of Alabama		General Land Office
		Department of Environmental Quality				Commission on Environmental Quality
		Department of Wildlife and Fisheries				
		Department of Natural Resources				

Source: Congressional Research Service based on data provided by Michael G. Jarvis, Congressional Affairs Specialist, NOAA.

Typically, Trustees work together, forming a Trustee Council, to develop a restoration plan that addresses all of the damages to all of the Trustees' resources. These Trustees must reach consensus on the extent of damages and restoration when issuing a unified plan. The statutory obligation of each Trustee is to "develop and implement a plan for the restoration, rehabilitation, replacement, or acquisition of the equivalent, of the natural resources under their trusteeship."[19] When the goal is to have one plan to address all of the impacts, which is way NOAA generally operates, the Trustees must work cooperatively to determine the magnitude and extent of injury to natural resources and create a plan to restore those injured resources to baseline (pre-spill) levels. Each state gets one vote on these issues, even if a state has multiple state agencies represented among the Trustees.

Do the Trustees Have to Work Together?

Past NRDA processes have occurred on a much smaller scale with fewer Trustees. The size of the 2010 spill and the mix of Trustees may make consensus among them more difficult. Trustees have an incentive to work under the NRDA process: courts have held that no litigation may be brought under OPA unless the regulatory process is completed.[20] But it is not clear that they have to work together to develop one unified plan.

The act does not appear to ban the possibility of multiple, separate NRDA processes from one spill. It states only that the act will not provide double compensation for the same loss.[21] Section 2706(c) assigns each type of Trustee (federal, state, tribal, and foreign) the responsibility of developing its plan for the restoration of the resources it oversees, rather than requiring all the Trustees to develop just one plan for all damaged resources. This suggests that many plans would be allowed under OPA. To the extent that the damage can be cleanly divided among Trustees, this may not be problematic. However, natural resources frequently do not have political boundaries, and it is possible that different Trustees may argue the same resources belong to them. A situation where the Trustees were each acting separately could lead to a bidding war for settlement with the responsible parties. One Trustee could develop a plan for resources and obtain compensation before another Trustee could develop its plan for the same resources.

The NOAA regulations contemplate a separate process, although most of those regulations are written to describe a process where multiple Trustees create one, unified plan. The regulations provide that the Trustees may act separately where the resources can reasonably be divided.[22]

Congress identified these issues and recognized that separate plans may result, while indicating that cooperation was the preferred method. After acknowledging that in some cases more than one Trustee may share control over a natural resource, the House Conference Report on OPA states that "trustees should exercise joint management or control over the shared resources.... The trustees should coordinate their assessments and the development of restoration plans, but [OPA] does not preclude different trustees from conducting parallel assessments and developing individual plans."[23]

Covered Natural Resources

The natural resources typically covered by NRDA include air, water (including ground water), soil, sediment, ocean bottom, biota (including bird, fish, and invertebrates), and habitat (for example, marshes, mangroves, mudflats, and vegetation). Of particular concern for the Gulf NRDA process are: marine mammals and sea turtles, fish and shellfish, birds, deep water habitat (for example, deepwater corals and chemosynthetic communities), intertidal and near shore subtidal habitats (including sea grasses, mud flats, oyster beds, and coral reefs), shoreline habitats (including salt marshes, beaches, and mangroves), terrestrial wildlife, and habitats (for example, alligators and terrapins). A useful discussion of the species and habitat at risk from the oil spill is available in CRS Report R41311, *The Deepwater Horizon Oil Spill: Coastal Wetland and Wildlife Impacts and Response*, by M. Lynne Corn and Claudia Copeland.

Management responsibilities for all natural resources within state territories fall to the states, except for specific resources for which the federal government has assumed responsibility by statute. For example, the Endangered Species Act, the Marine Mammal Protection Act, and the Migratory Bird Treaty Act all give the federal government control over the animals they cover. Wildlife not covered by federal statute is under state control. Waters and lands within state territory are also under state control. Waters protected by the National Marine Sanctuaries Act are under federal control, as are any federal lands such as those within the National Wildlife Refuge System, or National Parks, National Seashores, or National Recreation Areas. The federal government is also responsible for all resources beyond state territorial waters (usually three miles from shore).[24]

NRDA also contemplates how people enjoy common resources, but it does not compensate for individual losses—only the Trustees may collect. The services the natural resources provide, such as recreational fishing, boating, and shoreline recreation, may also be considered in the NRDA process.[25] For example, marshes serve as a buffer from hurricanes and fish provide a fishery to humans. However, NRDA money can only be used to restore the marsh or fishery, not to reimburse people whose houses are damaged by a hurricane or fishermen who are unable to earn a living from fishing. An exception is provided for subsistence use of resources, which is counted as a type of compensatory damage under OPA.[26]

Responsible Parties

The parties responsible for causing the oil spill will be responsible for NRDA damages. In the case of offshore drilling, a *responsible party* is the lessee or permittee of the area in which the facility is located.[27] Soon after the spill, the Coast Guard must designate the responsible parties.[28] The Coast Guard notified BP it was a responsible party for the spill on April 28, 2010.[29] The Trustees must give a written invitation to the responsible parties to participate in the NRDA process, and if the responsible parties accept, they must do so in writing.[30]

OPA imposes joint, several, and strict liability.[31] *Joint and several liability* means that where there are multiple responsible parties, each is potentially liable for the whole amount of the damages, regardless of its share of blame. (A separate action for subrogation could be brought by responsible parties to sort out reimbursement issues.[32]) *Strict liability* means liability is assigned regardless of fault or blame. There does not have to be a mistake, negligence, or a willful action for a party to be responsible.

Determination of Damages

OPA states that responsible parties are liable for "removal costs and damages" that result from an incident.[33] *Removal* is defined by the regulations to be synonymous with *response*.[34] Response includes containing and removing oil, and other actions to minimize and mitigate damage.[35] Three measures for calculating damages are authorized by 33 U.S.C. § 2706(d). The first allows "the cost of restoring, rehabilitating, replacing, or acquiring the equivalent of, the damaged natural resources." The second takes into account "the diminution in value of those natural resources pending restoration." And the third allows recovery for those costs incurred in "assessing those damages." Damages are capped under OPA unless an exception applies. For offshore facilities, a responsible party's liability for economic damages would end at $75 million, but would have no cap on removal costs.[36] Harm to natural resources is categorized as a *damage* under OPA; *removal* is separate.[37] Exceptions that would nullify the cap include gross negligence, willful misconduct, or violating an applicable federal regulation.[38]

The damages section of OPA also gives the Trustees a benefit should the matter advance to trial. Under Section 2706(e)(2), if the Trustees satisfy the regulatory requirements of OPA in estimating damages, their assessment is given a rebuttable presumption of accuracy in any hearing. This means that a responsible party would have the burden of proving that the assessment is wrong, rather than the Trustees having to show that the assessment is right.

OPA provides a federal remedy for recovery of damages. Different liability may be imposed under other laws, however. For example, criminal liability for harming protected species may still be pursued.[39] States may have their own laws.[40] The statute specifically allows states to impose additional liability for oil spills and/or requirements for removal activities.[41]

Once money is recovered for any natural resource purpose, including to cover the costs of assessing the damages, it is deposited in a special account for the express purpose of restoring Trustees' resources.[42]

HOW THE NRDA PROCESS WORKS

The NOAA regulations for OPA describe the Trustees' work as taking place in three steps: a Preassessment Phase, the Restoration Planning Phase, and the Restoration Implementation Phase.[43] These phases are discussed in detail below.

Preassessment Phase

Three main activities occur in the Preassessment Phase.[44] First, the Trustees establish whether there is jurisdiction under OPA and whether it is appropriate to try to restore the damaged resources. Under 15 C.F.R. § 990.42, the Trustees must determine that there are injuries, that those injuries have not been remedied, and that there are feasible restoration actions available to fix the injuries. If any of those evaluations result in a negative finding, the NRDA process ends. This step involves data gathering, and the Trustees use multiple sources, including the public, to obtain the information they need.

Once injuries have been found, the Trustees complete the second step of the Preassessment Phase—preparation of a Notice of Intent to Conduct Restoration Planning Activities. This Notice is published in the *Federal Register* and also is delivered directly to the responsible parties.

The third step for the Trustees in the initial phase is to open a publicly available administrative record. The record includes the documents considered by the Trustees throughout the process. The Federal Lead Trustee (Department of the Interior in this case) will choose the physical location(s) of the record. This record stays open until the Final Restoration Plan is delivered to the responsible parties.

Preassessment for the 2010 Oil Spill

The NRDA process in the Gulf is currently in the Preassessment Phase.[45] The Notice of Intent to Conduct Restoration Planning Activities is expected in September 2010, according to NOAA. Technical Working Groups composed of state and federal natural resource Trustees and representatives from BP's environmental consulting firm, Entrix, are gathering scientific information and are implementing baseline and post-impact field studies for multiple resource categories. Currently, the resources being assessed include marine mammals and sea turtles, fish and shellfish, birds, deep water habitat (deepwater corals and chemosynthetic communities), intertidal and near shore subtidal habitats (including sea grasses, mud flats, oyster beds, and coral reefs), shoreline habitats (beaches, salt marsh, mangroves), terrestrial wildlife and habitats, and human uses (recreational fishing, boating, and beach recreation). Samples of water, sediment, and tissues are being collected via land and ship-based sampling and aerial surveys. The Trustees will assess impacts from the response, including dispersant use at the surface and at depth.

Restoration Planning Phase

Phase two focuses on designing the restoration. During this phase, known as the Restoration Planning Phase,[46] Trustees quantify injuries and indentify possible restoration projects. In addition to identifying the nature of the harm to the resource from the oil spill, the Trustees will also evaluate harm resulting from the response actions,[47] such as the in situ burning, the use of dispersants, or vehicle damage to shores and marshes. These injuries are also compensable under OPA.[48]

Activities include field studies, data evaluation, modeling, injury assessment, and quantification of damage, either in terms of money needed to restore the resource or in terms of habitat or resource units. It is at this stage that the baseline is established. The baseline is the level the Trustees agree the resources were at prior to the injury and to which they will be restored under NRDA.[49] The regulations allow the Trustees to use historical data, reference data, control data, and/or data on incremental changes to establish the baseline.[50]

In practice, this has meant that where there are no baseline data for a certain species, the Trustees might look at a similar species to extrapolate data. It is not practical to expect to have up-to-date baseline data for every species everywhere there might be an oil spill. Instead NOAA has indicated that its practice is to identify the highest priority species and use this species as a proxy for those species for which data are not available. Another method is to establish a *guild* of species that have similar habitats, such as species of fish. Even if the impact on one species of fish in that guild is unknown, data may be gleaned for that species based on how the other fish are affected.

The goal of accumulating this information is to formulate a restoration plan that includes specific projects for remediation. This requires calculating the discounted values of the resources. Certain systems are in place from other NRDA events to help define the scope of the problem. For example, NOAA uses modeling and other procedures such as a Habitat Equivalency Analysis and Resource Equivalency Analysis to help quantify the scale of loss.

Before the restoration plan can be drafted, the Trustees assemble a panoply of restoration alternatives, which, for a cleanup on the scale of the 2010 oil spill, will include a broad range of projects directed at wildlife restoration, habitat restoration, and projects to provide for the loss of services and functions these resources provide. It is possible for the final projects to encompass five states, so the scope of the initial range will be considerable. The alternatives could include primary or compensatory restoration components, or both, provided they address specific injuries from the spill.[51] For an examination of the different types and methods of restoration, see "Restoration Options," below.

Once the range of alternatives is chosen from this list, the Trustees will evaluate the alternatives and choose one as the basis of the restoration plan.[52] The public is involved throughout the data gathering process. The Draft

Damage Assessment and Restoration Plan is submitted to the public for formal comment. Those comments are addressed within the Final Restoration Plan.

When the Final Restoration Plan involves federal resources, it must be reviewed under the National Environmental Policy Act (NEPA).[53] NEPA requires that major federal actions that significantly affect the human environment must be reviewed to learn the impacts of the action.[54] The extent of the environmental review depends on the extent of the impacts on the environment. Final Restoration Plans that have significant impacts on the human environment will result in an environmental impact statement, evaluating the impacts, alternatives to the chosen activity, possible mitigation, and involving the public in the process. Lesser impacts may mean that an environmental assessment is appropriate.

Restoration Implementation

Once the Trustees have agreed on a Final Restoration Plan, they move to phase three, Restoration Implementation.[55] The Final Plan is presented to the responsible parties, who have 90 days to respond. If the responsible parties agree to the plan, they may enter a settlement agreement with the Trustees. This agreement outlines what restoration work will be done, who will pay for it, and how damages discovered later will be handled. The responsible parties could undertake to perform the restoration activities on their own, they could pay for others to do the work, or both.

Where financial compensation, rather than primary restoration, is due, the responsible parties must agree to make the payments, although a schedule for funding could be negotiated.

If a responsible party does not agree to pay the damages or remediation expenses outlined in the Final Plan, the Trustees have two options. The Trustees may file suit in federal court or they may submit a claim for damages to the Oil Spill Liability Trust Fund. (See "Oil Spill Liability Trust Fund," below, for an examination of this account.) If the Trust Fund is used, the federal government is authorized to recover any compensation paid by the fund from a responsible party.

NRDA Funding for the 2010 Oil Spill

BP established a $20 billion escrow fund targeted towards individual and business losses from the oil spill.[56] This fund is known as the Gulf Coast Claims Facility, which went into operation August 23, 2010.[57] It is not a fund for government NRDA expenses, but it will provide for reimbursement for subsistence use losses of natural resources by individuals or businesses.

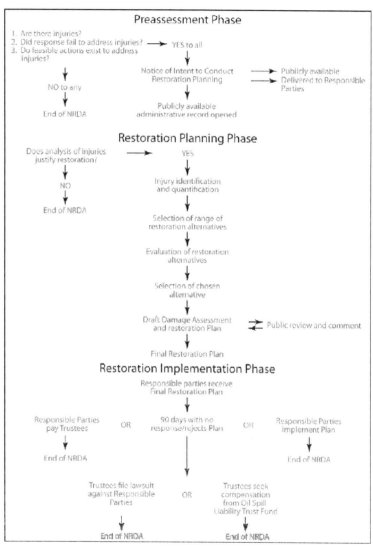

Source: Congressional Research Service based on 15 C.F.R. Part 990.
Note: Preassessment Phase—15 C.F.R. §§ 990.40-990.45; Restoration Planning Phase—15 C.F.R. §§ 990.50- 990.56; Restoration Implementation Phase—15 C.F.R. §§ 990.60-990.66.

Figure 1. Flow Chart of NRDA Process. According to NOAA Regulations.

MORE DETAILS ABOUT NRDA

Restoration Options

Restoration can include restoring, replacing, rehabilitating, or acquiring the equivalent of the natural resource harmed or destroyed by the incident.[58] Restoring the resource where the injury occurred is called primary restoration. Allowing the injured resource to recover naturally is a type of primary restoration. However, primary restoration is not always practicable, either via natural recovery or by human cleanup. When that is the case, compensatory restoration may be needed. Compensatory restoration is an action or payment to make up for the interim or permanent loss of a resource. For example, allowing a swamp that was oiled by the spill to recover on its own may be preferred, since oiled marsh is particularly difficult to clean without causing even more harm to the area. If the marsh is allowed to recover on its own, which could take decades, the Trustees could recover compensatory damages for the loss of benefits from that marsh until it returns to its baseline condition. The money paid for this interim period could be used to clean up an area damaged by some other cause or to enhance a similar marsh.

NOAA has indicated it prefers compensatory restoration to be *in kind*, that is, to enhance a marsh to make up for an oiled marsh which cannot be restored. In kind restoration, however, is not always feasible. There may be no parallel resource. In that case, NRDA permits restoration *out of kind*. For example, one unit of near-shore habitat might be found to have similar environmental benefits as one half unit of marsh. If the damaged near-shore habitat or another near-shore habitat cannot be restored (an *in kind* restoration), then NOAA could choose, for example, to restore one half unit of marsh for every one unit of damaged near-shore habitat (*out of kind* restoration). Habitat restoration typically occurs on publicly owned lands; however, out of kind restoration can occur on private lands, if that land provides habitats for injured animals, with the owner's permission.

The Trustees could find that replacing a natural resource, rather than restoring it, makes the most sense. Oyster beds are an example of where replacing a natural resource might be suitable: if the bed is totally destroyed, the bed might be replaced with new oysters. When the resource cannot be restored and there is no similar resource nearby to restore in its place, an *equivalent* resource could be acquired. For example, if a public beach were destroyed, the Trustees could buy a private beach and make it public by providing public access. According to the House Conference Report, the priority is "to restore, rehabilitate and replace damaged resources. The alternative of acquiring equivalent resources should be chosen only when the other alternatives are not possible, or when the cost of those alternatives would, in the judgment of the trustee, be grossly disproportionate to the value of the resources involved."[59]

Oil Spill Liability Trust Fund

Natural resource damages could be paid by the Oil Spill Liability Trust Fund (OS Trust Fund),[60] if the responsible parties refuse to accept the Final Restoration Plan and the Trustees choose not to sue. The OS Trust Fund is administered by the Coast Guard. It is financed chiefly by a per-barrel tax on crude oil produced in or imported to the United States.[61]

OS Trust Fund monies are available for a range of remedial and compensatory uses, even during the NRDA process.[62] For example, money for the Trustees' immediate assessment of the natural resource damage may come from the OS Trust Fund until the responsible parties are identified and provide reimbursement. The OS Trust Fund has limits for compensating for damaged natural resources. The OS Trust Fund could be used to pay the damages exceeding an offshore facility's liability limit under OPA ($75 million for economic damages[63]), up to its per-incident cap of $1 billion.[64] Only $500 million of that amount can go towards natural resource damages, however.[65] The OS Trust Fund could also be used if the responsible parties are not known, insolvent, or refuse to give money for assessment before they are found responsible by a court.[66]

Settlement vs. Litigation

According to NOAA, in the case of most spills, the Trustees and the responsible parties resolve the details of the restoration process via a settlement agreement. Settlement may occur at any time, provided that the terms of the settlement are adequate to satisfy the goal of OPA and are "fair, reasonable, and in the public interest."[67] Settling quickly after a spill may be desirable to the Trustees because the public is still engaged in the oil spill response. However, waiting longer before settling could allow a more reliable assessment of long-term effects on natural resources and a better calculation of the recovery costs.

A settlement agreement could include a provision requiring that assessment of the condition of the resources be conducted on a regular basis. A settlement could also include a reopener clause, like the one in the *Exxon Valdez* settlement. (The *Exxon Valdez* spill predated OPA, but the resolution of the resulting natural resources claims is useful as it is the closest in scale to the 2010 oil spill.) The purpose of a reopener clause is to provide a chance for Trustees to make claims years after settlement if they discover new damages to their resources from that original spill. Some have argued that the reopener clause in the Exxon settlement contained provisions that were not favorable to the Trustees.[68] For example, in order to make a claim under the reopener, any damage could not have been known or reasonably anticipated at the time of the settlement. Additionally, the *Exxon Valdez* agreement did not include any schedule for resolving claims brought under the reopener, allowing at least one claim to linger over five years with no sign of resolution.[69]

If settlement negotiations on the 2010 spill are unsuccessful, and the responsible parties reject the Final Restoration Plan or fail to respond within 90 days of receipt of the plan, the Trustees can file suit in federal court against the responsible parties under NRDA. NRDA claims must be filed within three years of the Final Restoration Plan.[70] At least one court has held that the responsible parties could demand a jury for the trial.[71] If the NRDA issues go to litigation, any discussions during the settlement negotiations are privileged. The court order resulting from a NRDA suit would likely outline the restoration process and who would pay for it. The responsible parties would then be legally bound to follow the order. BP may find some benefit in rejecting the Final Restoration Plan; it would delay any payment ultimately due until the court process was completed. (The final court case in the *Exxon Valdez* punitive damages was resolved by the U.S. Supreme Court 19 years after the spill,[72] although the company began paying natural resource damages under a settlement in the 1990s.) However, settlement offers BP the advantage of having some control over its fate—something a court case does not. Additionally, at trial BP would have the burden of *dis*proving the correctness of the Trustees' Final Restoration Plan.

Legislative Considerations

Congress has shown interest in Gulf restoration, although NRDA recovery under OPA has not been specifically addressed. The House passed H.R. 3534, which, in Title V, proposes the formation of a Gulf of Mexico Restoration

Program.[73] The program appears similar to the NRDA process: it would create a task force comprising the governors of Gulf states and heads of appropriate agencies. The task force would develop strategies for restoring natural resources in the Gulf and issue reports every five years. H.R. 3534 would also require the Secretary of the Interior to organize baseline studies for the Gulf region.[74] It appears that the process proposed in Title V of H.R. 3534 would overlap with NRDA. It is unclear if this is intended to preempt the NRDA process or provide a parallel, perhaps redundant, system. Section 701 of H.R. 3534 would eliminate the $75 million liability cap for offshore facilities.

Some issues relevant to NRDA before Congress are

- requiring better, ongoing baseline data collection for use in assessing future spills;
- allowing NRDA money to be used for research and development and advance planning of NRDA implementation;
- calling for long-term, comprehensive ecological studies of the effects of oil spills, e.g., 20 years later there is still oil on the beaches of Prince William Sound from the Exxon spill, and researchers continue to learn about the effects of the spill on various fish and birds;[75]
- codifying the terms of a potential settlement between BP and the Trustees, akin to the settlement relating to the San Joaquin River;[76]
- prohibiting responsible parties from "buying up" experts.[77]

CONCLUSION

The NRDA process has been successful in the past, but it has never been tested on such a large a scale as the 2010 Gulf oil spill. More oil was spilled, a greater geographic area is involved, and more Trustees are affected than in past spills. The Trustees may have difficulty agreeing on the assessment of damages, baseline conditions, the value of the damaged resources, and the proper method of restoring them. If a unified restoration plan is sought, the Trustees must make unanimous decisions on these issues, and then BP has the option not to accept the Final Restoration Plan. If BP rejects the Trustees' Plan, the Trustees may sue BP under NRDA for resolution of these issues, extending the final conclusion—restoration of the natural resources— even further.

ACKNOWLEDGMENTS

The author would like to thank Perry Cooper, a law student intern, who contributed to this report.

End Notes

[1] An estimated 4.9 million barrels (bbl) were released, but approximately 800,000 bbl were captured before spilling into the Gulf. See Official Site of the Deepwater Horizon Unified Command, at http://www.deepwaterhorizonresponse.com/ go/doc/2931/840475.
[2] 42 U.S.C. § 9607(a)(4)(C).
[3] 33 U.S.C. § 1321(f)(5).
[4] 33 U.S.C. § 2702(b)(2)(A).
[5] 16 U.S.C. § 1436.
[6] 16 U.S.C. § 19jj-1.
[7] The United States' exclusive economic zone extends to 200 nautical miles offshore; the *Deepwater Horizon* spill occurred 50 miles offshore. See 33 U.S.C. § 2701(6).
[8] See In re: Settoon Towing, No. 07-1263, 2009 WL 4730971 (E.D. La. Dec. 4, 2009); Gabarick v. Laurin Maritime (America) Inc., 623 F. Supp. 2d 741 (E.D. La. 2009).
[9] See Rice v. Harken Exploration, Inc., 250 F.3d 264, 266 (5th Cir. 1991). OPA section 1001(17) (33 U.S.C. § 2701(17)) declares that OPA's liability standard is the same as that in CWA section 311, the provision of that act addressing oil spills. CWA section 311, in turn, has been interpreted by courts to impose strict, joint and several, liability.
[10] 33 U.S.C. § 2702(b). For an overview of OPA liability in general, see CRS Report R41266, *Oil Pollution Act of 1990 (OPA): Liability of Responsible Parties*, by James E. Nichols.
[11] OPA § 2002(a), 33 U.S.C. § 1321 note.
[12] 33 U.S.C. § 2702(b)(2)(A).
[13] 33 U.S.C. § 2701(20).
[14] NOAA, Southeast Region home page for Damage Assessment, Remediation, and Restoration Program, http://www.darrp.noaa.gov/southeast/index.html.
[15] 15 C.F.R. § 990.11.
[16] 135 Cong. Rec. 26943 (Nov. 2, 1989).
[17] 33 U.S.C. §§ 2713(a) and (c). This requirement has been held to be jurisdictional. See Boca Ciega Hotel, Inc. v. Bouchard Transp. Co., 51 F.3d 235, 240 (11th Cir.1995); Russo v. M/T Dubai Star, No. C 09-05158 SI, 2010 WL 1753187 (N.D. Cal. April 29, 2010); Marathon Pipe Line Co. v. LaRoche Indus. Inc., 944 F. Supp. 476, 477 (E.D. La.1996); Johnson v. Colonial Pipeline Co., 830 F. Supp. 309, 311 (E.D. Va. 1993); Abundiz v. Explorer Pipeline Co., 2003 WL 23096018, at *5 (N.D. Tex. Nov. 25, 2003); Prairie Band of Potawatomi Indians v. Glacier Petroleum, Inc., No. Civ. A. 00-2165-CM, 2001 WL 584451 (D. Kan. May 2, 2001) (dismissing the complaint for failing to complete the requisite stages under OPA).
[18] NOAA, Deepwater Horizon Oil Spill, http://www.darrp.noaa.gov/southeast/deepwater_horizon/index.html (last visited July 30, 2010).
[19] 33 U.S.C. § 2706(c)(1)(C).

[20] See Boca Ciega Hotel, Inc. v. Bouchard Transp. Co., 51 F.3d 235, 240 (11th Cir.1995); Russo v. M/T Dubai Star, No. C 09-05158 SI, 2010 WL 1753187 (N.D. Cal. April 29, 2010); Marathon Pipe Line Co. v. LaRoche Indus. Inc., 944 F. Supp. 476, 477 (E.D. La.1996); Johnson v. Colonial Pipeline Co., 830 F. Supp. 309, 311 (E.D. Va. 1993); Abundiz v. Explorer Pipeline Co., 2003 WL 23096018, at *5 (N.D. Tex. Nov. 25, 2003); Prairie Band of Potawatomi Indians v. Glacier Petroleum, Inc., No. Civ. A. 00-2165-CM, 2001 WL 584451 (D. Kan. May 2, 2001) (dismissing the complaint for failing to complete the requisite stages under OPA).

[21] 33 U.S.C. § 2706(d)(3).

[22] 15 C.F.R. § 990.14(a)(2).

[23] H.Rept. 101-653, 1990 U.S.C.C.A.N. 779, 787 (1990).

[24] Submerged Lands Act, 43 U.S.C. § 1312. The territorial waters of Florida and Texas, however, extend to three nautical leagues (about nine miles) into the Gulf. United States v. Louisiana, Texas, Mississippi, and Alabama, 363 U.S. 1, 36-66 (1960) (Texas territorial waters); United States v. Florida, 363 U.S. 121, 129 (1960).

[25] NOAA, Deepwater Horizon Oil Spill, http://www.darrp.noaa.gov/southeast/deepwater_horizon/index.html (last visited July 30, 2010).

[26] 33 U.S.C. § 2702(b)(2)(C). *Subsistence use* is not defined within the act or regulations but is commonly defined as only that amount which is consumed by the harvester and family, and not for commercial benefit.

[27] *Responsible party* is further defined at 33 U.S.C. § 2701(32)(C).

[28] The authority of the President to designate the responsible party under 33 U.S.C. § 2714(a) was delegated to the Coast Guard via Executive Order in 1991. Exec. Order No. 12777 (56 Fed. Reg. 54757 (October 22, 1991)).

[29] Email communication with the author on August 26, 2010 from LTCR Thomas A. Shuler, U.S. Coast Guard Deputy Senate Liaison.

[30] 15 C.F.R. § 990.14(c)(1).

[31] In re: Settoon Towing, No. 07-1263, 2009 WL 4730971 (E.D. La. Dec. 4, 2009). S. Rep. 101-94, 1990 U.S.C.C.A.N. 722, 726 (1990) ("[this bill] explicitly extends strict, joint, and several liability for compensation of third party damages").

[32] 33 U.S.C. § 2702(d)(1)(B).

[33] 33 U.S.C. § 2702(a).

[34] 15 C.F.R. § 990.30.

[35] 33 U.S.C. § 2701(30).

[36] 33 U.S.C. § 2704(a)(3).

[37] 33 U.S.C. § 2702(b).

[38] 33 U.S.C. § 2704(c)(1).

[39] For an analysis of criminal laws related to wildlife harm, see CRS Report R41308, *The 2010 Oil Spill: Criminal Liability Under Wildlife Laws*, by Kristina Alexander.

[40] A Congressional Distribution Memorandum by CRS is available on Oil Spill Liability Statutes in the Gulf States. Contact the author for a copy.

[41] 33 U.S.C. § 2718(a).

[42] 33 U.S.C. § 2706(f).

[43] 15 C.F.R. § 990.12.

[44] 15 C.F.R. Subpart D.

[45] http://www.darrp.noaa.gov/southeast/deepwater_horizon/index.html.

[46] 15 C.F.R. Subpart E.

[47] 15 C.F.R. § 990.51(e).
[48] 33 U.S.C. § 2702(a).
[49] See 15 C.F.R. § 990.30.
[50] 15 C.F.R. § 990.30.
[51] 15 C.F.R. § 990.53(a)(2).
[52] 15 C.F.R. § 990.55.
[53] 42 U.S.C. § 4332.
[54] For more information about NEPA, please see CRS Report RS20621, *Overview of National Environmental Policy Act (NEPA) Requirements*, by Kristina Alexander.
[55] 15 C.F.R. Subpart F.
[56] See Gulf Coast Claims Facility, http://www.gulfcoastclaimsfacility.com.
[57] *Id.*
[58] 15 C.F.R. § 990.30.
[59] H.Rept. 101-653, 1990 U.S.C.C.A.N. 779, 786-787 (1990).
[60] 33 U.S.C. § 2712. The standards and procedural requirements for claims filed against the OS Trust Fund are set forth in the Coast Guard's OPA regulations. See 33 C.F.R. §§ 136.1-136.241.
[61] 26 U.S.C. § 4611.
[62] For more on the OPA claims process, see CRS Report R41262, *Deepwater Horizon Oil Spill: Selected Issues for Congress*, coordinated by Curry L. Hagerty and Jonathan L. Ramseur.
[63] 33 U.S.C. § 2704(a)(3).
[64] 26 U.S.C. § 9509(c)(2)(A)(i).
[65] 26 U.S.C. § 9509(c)(2)(A)(ii).
[66] 33 U.S.C. § 2712 (a).
[67] 15 C.F.R. § 990.25.
[68] *Assessing Natural Resource Damages Following the BP Deepwater Horizon Disaster: Hearing Before the Subcomm. on Wildlife and Water of the S. Comm. on Environment and Public Works*, 111th Cong. 6 (July 27, 2010) (written testimony of Stanley Senner, Ocean Conservancy).
[69] *Id.*
[70] 33 U.S.C. § 2717(f)(1)(B).
[71] United States v. Viking Resources, Inc., 607 F. Supp. 2d 808 (S.D. Tex. 2009).
[72] Exxon Shipping Co. v. Baker, 128 S. Ct. 2605 (2008).
[73] H.R. 3534, Tit. V (111th) (as passed by House, July 30, 2010).
[74] H.R. 3534, 111th Cong. § 224 (as passed by House, July 30, 2010).
[75] *Assessing Natural Resource Damages Following the BP Deepwater Horizon Disaster: Hearing Before the Subcomm. on Wildlife and Water of the S. Comm. on Environment and Public Works*, 111th Cong. 5 (2010) (written testimony of Stanley Senner, Ocean Conservancy).
[76] P.L. 111-11, Tit. X.
[77] Mark Tran, *BP Denies "Buying Silence" of Oil Spill Scientists*, The Guardian (July 23, 2010) available at http://www.guardian

In: Gulf Oil Spill of 2010…
Editors: C. R. Walsh, J. P. Duncan

ISBN: 978-1-61324-729-7
© 2012 Nova Science Publishers, Inc.

Chapter 5

NATURAL RESOURCE DAMAGE ASSESSMENT: EVOLUTION, CURRENT PRACTICE, AND PRELIMINARY FINDINGS RELATED TO THE *DEEPWATER HORIZON* OIL SPILL[*]

National Commission on the BP Deepwater Horizon Oil Spill and Offshore Drilling

STAFF WORKING PAPER NO. 17

Staff Working Papers are written by the staff of the BP Deepwater Horizon Oil Spill Commission for the use of the members of the Commission. They do not necessarily reflect the views of the Commission as a whole or any of its members. In addition, they may be based in part on confidential interviews with government and non-government personnel.

Six months after the oil has stopped flowing from BP's damaged Macondo well, the amount of environmental harm caused by the spill is uncertain, as is the adequacy of existing legal, regulatory, and policy mechanisms to ensure that restoration needed to redress the damage will be

---

[*] This is an edited, reformatted and augmented version of a National Commission on the BP Deepwater Horizon Oil Spill and Offshore Drilling publication, Staff Working Paper No. 17.

fully implemented by government and paid for by responsible parties. This background paper describes the process that was established under the Oil Pollution Act of 1990 for assessing natural resource damages caused by the spill and restoring damaged resources to their pre-spill condition. Known as Natural Resource Damage Assessment (NRDA), this process is still in the early phases of being applied to the BP spill and conclusions about its efficacy or success in this instance will be impossible to draw for a number of years, possibly decades. This background paper describes the history and purpose of the NRDA, reviews the main steps in the NRDA process, and reports on the status of current damage assessment efforts in the Gulf.

NATURAL RESOURCE DAMAGE ASSESSMENT: HISTORY AND PURPOSE

In the wake of the *Exxon Valdez* disaster in 1989, Congress passed legislation specifically aimed at responding to and addressing environmental and economic damages from oil spills. As part of the Oil Pollution Act of 1990 (OPA), 33 U.S.C. §§ 2701 et seq., "responsible parties"[1] were made liable for the removal costs and damages resulting from discharges of oil from vessels or facilities. Among other things, this liability extends to:

> Damages for injury to, destruction of, loss of, or loss of use of, natural resources, including the reasonable costs of assessing the damage, which shall be recoverable by a United States trustee, a State trustee, an Indian tribe trustee, or a foreign trustee.[2]

The measure of damages under OPA is:

a) The cost of restoring, rehabilitating, replacing or acquiring the equivalent of, the damaged natural resources;
b) The diminution in value of those natural resources pending restoration; plus
c) The reasonable cost of assessing those damages.[3]

Under OPA, responsibility for promulgating regulations to guide the assessment of natural resource damages fell to the National Atmospheric and Oceanic Administration (NOAA).[4] NOAA completed this task in 1996 and NRDA regulations became effective on February 5, 1996.[5]

Prior to 1990, natural resource damage assessments and the associated cost recovery for oil spills were governed by the Comprehensive Environmental Response, Compensation and Liability Act of 1980, or CERCLA, which imposes liability for damages resulting from releases of "hazardous substances" as defined by the statute.[6] CERCLA regulations provided the model for the natural resource damage assessment authority set forth in OPA and continued to govern damage assessments for oil spills between 1990 and 1996, when NOAA was developing new regulations under OPA.

In its OPA regulations, NOAA seeks to promote cooperation between the trustees and the responsible party in carrying out the natural resource damage assessment. This process, referred to as a cooperative assessment, is being used in the *Deepwater Horizon* case, where BP (a "responsible party"[7]) is working with government agencies (i.e., the "trustees") to identify and quantify damages.[8] NOAA guidance documents set forth the specifics of the cooperative process, including level of participation, dispute resolution, agreement on scientific methods, sharing of equipment and experts, and funding.[9] As the guidance suggests, these issues are generally laid out in a memorandum of agreement between the trustees and the responsible party.[10] While past attempts to use the cooperative assessment process did not measurably shorten the time or administrative costs incurred between the event and final settlement, trustees are quick to point out that the cooperative assessment process provides other advantages.[11] In particular, states that do not have dedicated natural resource damage assessment programs[12] maintain that they would not have the budget or resources to carry out damage assessments if not for the cooperative agreement that allows for periodic funding and the sharing of equipment and experts. Further, one trustee also pointed out that the more recent emphasis on data collection – as opposed to making assumptions based on past spills or existing scientific knowledge – to quantify damages and reach settlement has lengthened the time to settlement but strengthened the process and its outcome.

NATURAL RESOURCE DAMAGE ASSESSMENT VERSUS CIVIL AND CRIMINAL PENALTIES

It is important to highlight the distinction between legal action to recover costs for damages to natural resources, and enforcement actions aimed at imposing civil or criminal penalties on the responsible party under an environmental statute. Both actions may be pursued, under separate authority, by states and the federal government in response to an event such as an oil spill. In bringing an enforcement action for civil or criminal penalties, the Department of Justice – on behalf of EPA, the Coast Guard, or another agency – acts in the role of prosecutor. By contrast, when the Department of Justice sues to recover natural resource damages, it is acting on behalf of the trustees with jurisdiction over the injured resources and the action is in many ways similar to a tort action.[13] As a general rule, funds recovered as a result of civil or criminal enforcement actions under federal environmental statutes are deposited in the federal treasury and may not be used to redress the harms caused by the pollution event or incident.[14] The authority to recover costs for damages to natural resources, conversely, is unique in that the funds recovered from responsible parties must be used to restore the specific resources injured by the event.[15]

UNDERSTANDING THE NRDA PROCESS

NRDA is the regulatory process used by designated natural resource trustees to identify, assess and restore damages to: (1) public natural resources, (2) the ecosystem services they provide (e.g. oysters provide water filtration) and (3) the public's lost use of those resources. Based on the damage assessment, the trustees either bring a lawsuit against the responsible party to recover the damages (which may be settled), or enter into a settlement with the responsible party without filing a lawsuit.

When an oil spill occurs, the trustees must work through three phases to determine the appropriate type and amount of restoration required to compensate the public:

1) *Preliminary Assessment* (also referred to as pre-assessment). In the aftermath of the release, the trustees collect time-sensitive data and observations and conduct research to determine if damage to a particular resource has occurred or is likely to occur. Did damage likely occur? If so, the trustees move to the next phase.
2) *Restoration Planning* (including injury assessment). In this phase, the trustees conduct scientific and economic studies to quantify damages and use local knowledge and expertise to identify potential restoration projects. A draft restoration plan describing potential compensatory

restoration projects and recommending preferred projects based on applicable regulatory criteria is made available to the public for review and comment. The National Environmental Policy Act requires that any potentially significant environmental impacts of the proposed restoration activities be considering during the process of reviewing the draft plan.[16]

3) *Restoration Implementation.* At this point, the restoration plan as proposed by the trustees and reviewed by the public is implemented and monitored to ensure its success.

Though the logic of this progression is straightforward, its implementation is anything but. Identifying and quantifying damages, particularly where complex ecosystems are involved, presents enormous challenges. Developing sound sampling protocols that cover adequate time scales, teasing out other environmental disturbances, and scaling the damages to the appropriate restoration project often takes considerable time; in fact, a typical damage assessment can take years. Two sets of determinations – one concerning the baseline conditions against which damages will be assessed and the second concerning the quantification of those damages – are particularly difficult and consequential in terms of the overall assessment results.

Determining Baseline Conditions

Natural resource damages must be measured against baseline conditions. The Oil Pollution Act defines the baseline against which damages are to be measured as "... the condition of the natural resources and services that would have existed *had the incident not occurred*."[17] Making this determination, however, is inherently difficult and often highly contentious. Baseline conditions may be estimated, according to the OPA regulations, "using historical data, reference data, control data, or data on incremental changes (e.g., number of dead animals), alone or in combination, as appropriate."[18] Without a well-established and agreed-upon definition of baseline conditions, there can be no agreement about a subsequent assessment of damages or quantification of required restoration. Given that the ecological baseline can vary seasonally, annually, and over much longer time scales, it can be difficult for all parties to agree upon the exact condition of an ecosystem prior a spill. Since long-term historical data sets are often non-existent or discontinuous in many areas of the country, natural resource trustees are likely to be disadvantaged by a lack of sufficient data to fully characterize the condition and trends of relevant ecosystems prior to the incident in question.

Quantifying Damages

Once baseline conditions have been established, an effort is made to quantify damages. Notwithstanding its inherent difficulty, quantification is needed to determine the appropriate amount of restoration required to compensate for the natural resource damages that have been incurred. Scientists use various methods to measure a reduction in ecological resources and the services they provide, as well as, the public's reduced enjoyment of the resources. These methods are highly dependent on the resource being assessed and on the proxies available for measuring the ecological function. For example, one study may use measured reductions in nutrient filtration to determine relative impact on a wetland while another may use decreased fish populations in future years to determine the relative impact the spill may have had on juvenile fish or larvae. To measure the public's reduced enjoyment of a resource, say recreational boating, a survey might to conducted to determine if the spill forced boaters to change plans, travel further, or forego trips altogether.

Typically, there is a fluctuating ecological baseline at the time of the spill. Once the spill occurs there is some decrease in the function of the resources (e.g., decrease in nutrient filtration, decline in an animal population, or loss of boating opportunities). If the ecosystem is left to recover naturally, it may eventually return to baseline conditions.[19] However, restoration efforts must be implemented to compensate for the damages that occur (relative to baseline) during this period of natural recovery.

There are three types of restoration options: emergency restoration, primary restoration and compensatory restoration.

> *Emergency Restoration.* Emergency restoration is invoked when time is of the essence to curtail a threat or reverse the damage to a particularly sensitive resource. For instance, emergency restoration would include the immediate planting of seagrass before the planting season ends for the purpose of mending damage left by response vessels. Time would be of the essence in that circumstance because the damage may worsen while waiting for the next planting season.
>
> *Primary Restoration.* Primary restoration is intended to return a damaged natural resource to baseline after an oil spill. For instance, if a

spill is killing marsh grasses along a wetland edge and exacerbating marsh erosion, the trustees may opt to plant new vegetation at the site of impact to curtail or reverse the inevitable land loss.

Compensatory Restoration. Once emergency and/or primary restoration (if implemented) and cleanup activities are completed, the trustees then quantify the remaining, or interim, natural resource damage when compared to baseline. In the end, the amount of compensatory restoration required should equal the full amount of interim natural resource damages incurred. For example, if 5000 adults birds were killed due to the spill, the responsible party must not only implement restoration to replace 5000 adult birds (e.g., buying prairie pothole land in the Midwest to dedicate to nesting habitat), it must also provide for the "production foregone," or the impact on future generations due to the loss of the adults (which would ultimately increase the number of animals required to be restored). With compensatory restoration, there is also a matter of compensating for the period of time for which the public has lost the resource. For simplicity, assume that 500 acres of marsh is lost in 2010 due to the spill. If restoration to restore the 500 acres is not completed until 2015, then the responsible party must pay interest, in the form of additional marsh acres, on the five years that the public did not have those 500 acres. The traditional interest rate in natural resource damage assessment is three percent.

APPLYING THE NRDA PROCESS TO THE DEEPWATER HORIZON OIL SPILL

Initial Response and Organization of Damage Assessment Activities

When the *Deepwater Horizon* explosion occurred on April 20, 2010, NOAA's Assessment and Restoration Division was already extremely busy conducting a number of oil spill and waste damage assessments and training activities. Ironically, the Division had just participated in a drill aimed at testing preparations for a "spill of national significance" (the drill took place in Portland, Maine in late March 2010). While such a spill had never been declared before, the Division was focused on evaluating and developing lessons learned from the drill, in coordination with its sister division, the NOAA Emergency Response Division, participating co-trustees, and the acting responsible party and sponsor for the drill, Shell Oil Company.

The Assessment and Restoration Division was first notified of the *Deepwater Horizon* explosion via a hotline report generated by NOAA's Chief Scientific Support Coordinator, Charlie Henry, in early morning hours of April 21, 2010. The initial report indicated that there were fatalities, there was oil on board the rig, and the rig was burning. At this point, the standard procedure for the Division is to stand by and initiate contact with other federal and potentially affected state and tribal trustees. By April 25, the Assessment and Restoration Division was on-scene in Houma, Louisiana and ready to begin collecting time-sensitive data that would help establish the toxicity of the oil and the baseline condition of potentially affected resources. By late April, the trustees were holding daily conference calls to provide situational updates and to construct and adapt future field sampling plans. By Saturday, May 1, experienced environmental and contaminant scientists from Florida, Alabama, and Mississippi, along with various academic institutions, began collecting coastal baseline data in anticipation of the oil reaching their respective coastlines. In Houma, NOAA, the State of Louisiana, and the Department of the Interior began organizing technical working groups and collecting baseline data along the Louisiana coast. Texas, with an experienced damage assessment program in place, was also engaged and was monitoring the movement of oil from the spill. The delay between the day of the explosion and the day the oil finally reached the coastline allowed the trustees to organize and collect vital baseline data over a large portion of the Gulf coastline and intermediate waters. Figure 1 lists the trustee agencies currently involved in the *Deepwater Horizon* damage assessment.

In the days following the *Deepwater Horizon* explosion, it became clear to NOAA's Assessment and Restoration Division that this was a generational spill that would require the majority of Division staff and resources to conduct a comprehensive damage assessment. State agencies with ad hoc natural resource damage assessment programs scrambled to provide staff that would be dedicated to the spill long-term. Two week rotational staffing assignments were quickly put in place and emergency contracts for technical support staff were activated. Charlie Henry, NOAA's Chief Scientific Support Coordinator for the spill warned that the response and impact assessment of this release would not be a sprint; it would be a marathon.

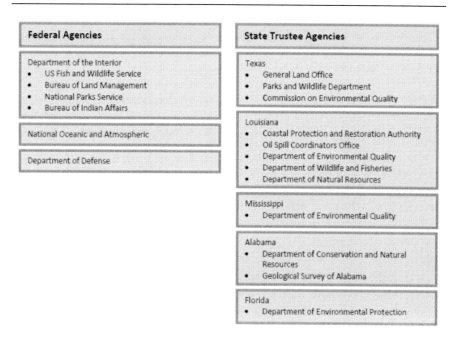

Figure 1. With two federal and 12 state agencies participating in the *Deepwater Horizon* NRDA process, coordinating schedules, reviewing documents, and communicating effectively across so many trustee agencies is complicated.

By early May, BP and the trustees agreed to work under the cooperative assessment framework to assess potential damages from the spill. Whether this process will work as intended under the OPA guidance remains uncertain at this time. In September, the trustees requested that BP fund several emergency and/or primary restoration projects to curtail impacts to migratory birds, submerged aquatic vegetation and wetlands. BP agreed to fund these projects and a migratory bird emergency restoration project has begun in Mississippi.[20]

At present, damage assessment activities being carried out across the Gulf are being managed from a central "war room" at the Incident Command Center in Houma, Louisiana. Needless to say, the scale of the undertaking represents new challenges for everyone involved – not only in terms of the geographic scale of the area being studied (both horizontally and vertically), but also in terms of the sophistication of the oceanographic equipment and the breadth of expertise being tapped to conduct the assessment. Scientists are observing oiled shorelines, tracking marine mammals, assessing fisheries impacts, and collecting water, oil, and sediment samples. BP and the trustees also have a handful of dedicated research vessels collecting data throughout the water column at various intervals.

Status and Early Results of the Damage Assessment Effort

As the language of the OPA regulations indicate, "baseline" for purposes of damage assessment in the NRDA context is generally considered to be the condition of the resource just prior to the spill. The precise application of this definition has particular importance in the Gulf of Mexico context, where many coastal habitats have been substantially degraded over decades – even centuries – under the pressure of ever-expanding industrial, commercial, and residential development. The NRDA regulations, as generally applied, require that BP restore Gulf resources to their functioning level as of April 19, 2010. However, the Gulf ecosystem in April 2010 was already weakened. Every year in the Gulf, for example, nutrient runoff from farms throughout the Mississippi River watershed creates a "dead zone" of extremely low oxygen levels in which few water organisms can survive. In some years, the area affected by this dead zone is as large as New Jersey. Throughout the region, erosion and destabilization of wetlands has been accelerated by the patchwork of canals carved out by the oil and gas industry. Cut off from natural deposits of sediment from the Mississippi River, delta wetlands have been unable to keep pace with rising sea levels and are sinking into the Gulf. These are only a few of the factors contributing to an imbalanced and already degraded Gulf ecosystem. In this context, effective long-term restoration will require the stabilization and eventual reversal of a number of long-standing, damaging trends.

As a first step in assessing damages from the *Deepwater Horizon* spill, state and federal trustees identified numerous categories of resources that might be at risk of adverse impacts, and began developing and carrying out preliminary assessment plans. Table 1 lists the specific resources being studied by the trustees and BP through their technical representative, Entrix, as part of the damage assessment process.

Table 1. The trustees are currently assessing potential damages to the following resource. If there is a determination that an injury occurred the magnitude of the injury will be quantified as part of the restoration planning phase

Resource Focus	Studies	Status
All Resources	• *Review historical information* to help document pre-spill conditions.	• *Ongoing*
Water Column and Sediment *Water Oil Sediment*	Document the amount of oil in the water, and determine how and where the oil is moving. • Various types of *Water quality surveys* document the presence of oil at various depths. • *Transect surveys and sentinel stations* detect submerged oil. • *Plume modeling and other studies* provide detail about the type of oil and how it moves in water. • *Sediment sampling* documents the presence of oil across habitats.	• *Ongoing* • *Ongoing* • *Ongoing* • *Ongoing*
Shorelines *Beaches Wetlands Mudflats Mangroves*	Document the extent and amount of oil on shoreline habitats. • *Aerial surveys* provide a bird's eye view of coastlines to determine the extent of oil; the resulting maps and data help target ground surveys. • *Ground surveys* allow scientists to collect more detailed data on the degree of oiling (e.g. light vs. heavy) and focus future data collection efforts.	• *Ongoing* • *Ongoing*
Aquatic Vegetation *Seagrasses Sargassum*	Document the presence/diversity of aquatic vegetation, and determine if it has been oiled. • *Aerial surveys* help identify where and to what extent aquatic vegetation may be oiled. • *Ground surveys* help identify location and extent of oiled aquatic vegetation.	• *Ongoing* • *Ongoing*

Table 1. (Continued).

Resource Focus	Studies	Status
Fisheries *Plankton* *Fish larvae* *Nearshore fish* *Offshore fish*	Document the presence/diversity of fish and plankton, and determine if they have been oiled. • *Plankton, invertebrate, fish, and fish larvae surveys* help determine the presence and/or abundance of these resources in oiled and non-oiled open water areas.	• *Ongoing*
Shellfish *Oysters Mussels* *Shrimp Crabs*	Document the presence/diversity of shellfish, and determine if they have been oiled. • *Oyster surveys* document the presence and/or abundance of oysters in affected areas, and provide scientists with tissue for lab analysis. • *Mussel collections* at monitoring stations help identify if mussels have been oiled, and if so, provide data for future studies. • *Shrimp collections* help document the presence and abundance of shrimp in the open water and in oil plumes.	• *Ongoing* • *Ongoing* • *Ongoing*
Corals *Shallow water corals* *Deep water corals*	Document the presence/diversity of corals, and determine if they have been oiled. • *Shallow-water coral surveys and tissue collection* help identify and evaluate exposure to existing communities. • *Deep-water coral surveys and tissue collection* help identify and evaluate exposure to existing communities. • *Monitoring devices* are installed in coral communities to determine exposure to oil.	• *Ongoing* • *Ongoing* • *Ongoing*

Resource Focus	Studies	Status
Marine Mammals and Turtles *Whales* *Manatees Dolphins* *Sea Turtles*	Document the presence/diversity of marine mammals and turtles, and determine if they have been oiled. • *Aerial surveys* document the location of marine mammals and turtles before they have been impacted by oil, and document the location and number of marine mammals and turtles that may be oiled, distressed, or dead; these surveys also document the potential changes in marine mammal behavior and distribution. • *Tissue sampling* from live and dead sea turtles and marine mammals helps assess oil exposure. • *Acoustic technology and satellite tags* help scientists assess the behavior and movement of marine mammals.	• *Ongoing* • *Ongoing* • *Ongoing*
Birds *Shorebirds Colonial seabirds* *Pelagic seabirds* *Secretive/marsh birds*	Document the presence/diversity of birds, and determine if they have been oiled. • *Ground surveys* identify injured, dead, or oiled birds on shorelines. • *Aerial and photograph surveys of open sea, shorelines, and islands* help identify the location and abundance of birds, and determine if they and/or their habitats have been oiled. • *Ground and boat surveys* in marshes document the abundance and degree of oil affecting marsh birds; *radio transmitters* provide for the assessment of bird movement and mortality. • *Point and transect boat surveys* help scientists monitor pelagic birds.	• *Ongoing* • *Ongoing* • *Ongoing* • *Ongoing*
Terrestrial Species *Terrapins* *Crocodiles* *Small Mammals*	Document the presence/diversity of terrestrial species, and determine if they have been oiled. • *Ground surveys* help identify and quantify oiled animals and/or habitats.	• *Ongoing*

Table 1. (Continued).

Resource Focus	Studies	Status
Human Use *Public beaches and parks* *Public facilities* *Cultural uses*	Document the many ways humans recreationally use and enjoy the natural resources of the Gulf, if these uses or enjoyment have been impacted by the spill, and if so, to what extent. • *Overflight Surveys* identify public beach use. • *Intercept Surveys* identify public boat ramp use. • *Information Surveys* to assess cultural uses.	• *Ongoing* • *Ongoing* • *Ongoing*

On September 29, the Trustees announced that they had found sufficient evidence of natural resource damage to file a Notice of Intent to Conduct Restoration Planning. This legal document signals a move from pre-assessment activities to "identify and document impacts to the Gulf's natural resources, and the public's loss of use and enjoyment of these resources, as the first stage under the regulations for developing a restoration strategy."[21]

As of mid-January 2011 over 28,000 water, sediment, tissue and tar ball samples have been collected and over 52,000 analyses completed.[22] These samples were collected during 89 off-shore research cruises and along approximately 4,215 linear miles of coastal shoreline.[23]

As of November 1, 2010 wildlife responders had collected 8,183 birds, 1,144 sea turtles and 109 marine mammals, alive and dead; oiled and unoiled. These numbers may have increased since then, but presumably, because the oil stopped flowing several months ago, the numbers will soon plateau if they have not already done so. Given that collection efforts are bound to miss some number of affected animals, many of which will never be found because of the effects of hiding, scavenging, sinking, decomposition, or the sheer size of the search area, the trustees will have to make assumptions to quantify impacts on wildlife. A common practice is to assign a multiplier to the final *observed* number of affected animals. The multiplier will vary depending on the species, its behavior, and its habitat. The multiplier is then used to estimate the *total* number of animals impacted.

Results of the assessment effort to date indicate that more than 650 miles of Gulf coastal habitats—salt marsh, mudflat, mangroves, and sand beaches—were oiled; more than 130 miles have been designated as moderately to heavily oiled. Oiled birds and beaches are often the most visually disturbing and widely disseminated images associated with a major oil spill, however, public and scientific concern in the *Deepwater Horizon* case has for some time focused on the impacts of an invisible sub-surface "plume" or "cloud" of oil. As part of the response and damage assessment effort being coordinated by BP and the trustees, 23 research vessels have been working to collect thousands of data points over 5000 feet of water column to assess potential impacts on subsurface biota, both from the oil and from the use of dispersants.

While the biological impacts are not fully yet understood, the National Incident Command's Joint Analysis Group, an inter-agency workgroup that was set up to analyze sub-surface data collected by scientists from federal, private, and academic institutions, released a report that described the chemical behavior of the subsurface oil.[24] The report summarizes 419 data points collected from 9 different vessels between May 8, 2010 and August 9,

2010. According to these data, depressed oxygen levels have been detected more than 80 km from the wellhead. The report concludes that while oxygen levels are depressed in the subsurface plume as a result of biodegradation (referred to in the report as biochemical oxygen demand), oxygen levels that would be detrimental to water column organisms have not been found and are not expected.

The Contribution from Non-Governmental Scientists

Three peer-reviewed studies have been published to date in *Science* related to the behavior of oil from the *Deepwater Horizon* spill in the deep sea environment. Another publication focuses on the potential toxicity of oil in the deep sea. Camilli et al. discovered, tracked and sampled a deepwater plume of dispersed hydrocarbons measuring at least 35 km long by 2 km wide and 200 m high at a depth of about 1100 m below the ocean surface from June 19 to June 28, 2010.[25] Interested, in part, as to whether microbial degradation of the plume would result in lethal depletion of oxygen in the water column[26] – or a "hypoxic zone" – the research team found no significant drawdown of oxygen inside the plume.

However, they do note that a hypoxic zone could ultimately develop over a period of time, as the microbes continue to degrade the oil therein: "[I]f the hydrocarbons are indeed susceptible to biodegradation, then it may require many months before microbes significantly attenuate the hydrocarbon plume to the point that oxygen minimum zones develop that are intense enough to threaten Gulf fisheries."[27]

Hazen et al. measured physical, chemical and microbiological properties of water samples taken from the same research area as Camilli, et al. from May 25 to June 2, 2010.[28] They report similar findings of only slight oxygen drawdown, and contend that the rate of biodegradation inside the plume is much faster than reported by Camilli et al.

Valentine et al. investigated dissolved hydrocarbon gases (methane, ethane, and propane) in the Gulf of Mexico water column from June 11 to 21, 2010. 29 This study again confirms the presence of the southwest plume at an average depth of 1100 m and identifies additional plumes, defined by elevated levels of methane, to the north and east of the well head, which probably were formed earlier when currents flowed in a different direction. The study suggests that the microbes in the plume have a preference for the lighter petroleum constituents (ethane and propane). They conclude, therefore, that the aging plume, once devoid of the lighter constituents, have bacterial populations that are primed for degradation of other hydrocarbons, but at a slower rate.

As of mid-January, one scientific study on the potential toxicity of this deep sea plume has been published. 30 Water samples collected from the discovered plume (discussed above) in mid-May contained levels of polycyclic aromatic hydrocarbon (PAH) that are high enough to be acutely toxic for some marine organisms. The PAH levels decreased with distance from the wellhead and persisted as far as 13 km from the source.

Taken together, these studies show the presence of deepwater plumes of highly dispersed oil droplets and dissolved gases between at 1000 and 1300 meters deep. Bacterial decomposition begins quickly for the light hydrocarbon gases propane and ethane but more slowly for the heavier hydrocarbons typically present in a liquid form and for the predominant gas, methane. The degradation rates are sufficient to reduce the dissolved oxygen concentrations, but not to harmfully low levels associated with "dead zones." Subsequent dilution with well-oxygenated, uncontaminated waters is sufficient to prevent any further drawn down of dissolved oxygen in the aging plumes. While oxygen levels do not appear to be lethal and the plume will dissipate with time, the PAH levels that accumulated in the deep sea plume likely had adverse and potentially acute effects on marine organisms that need to be further assessed and quantified.

A Historical Comparison

Human nature cannot resist the temptation for disaster comparison. Hence, the question is often asked, how does this spill compare to a previous oil spill volume record holder, the *Exxon Valdez* oil spill? At this point, because data are still being collected on water column and fisheries impacts, it is too soon to tell whether the immediate effects of the *Deepwater Horizon* oil spill on coastal areas and wildlife will turn out to be smaller in scale than those associated with the *Exxon Valdez* oil spill. Based on current information on marine life fatalities, that remains a possibility. In the aftermath of the *Exxon Valdez*, for example, more than 35,000 dead birds and 1,000 dead sea otters were recovered. Additionally, that spill oiled 1500 miles of Alaska coastline, of which 350 miles were heavily oiled. Those *Exxon Valdez* numbers are higher than currently known numbers for the *Deepwater Horizon* spill. At the end of their assessment, the *Exxon Valdez* Trustee Council estimated the final wildlife death toll to be "250,000 seabirds, 2,800 sea otters, 300 harbor seals, 250 bald eagles, up to 22 killer whales, and billions of salmon and herring eggs."[31] Because, however, the *Deepwater Horizon* spill has not been fully assessed and was of a very different character, most notably occurring in the subsea, a focus on readily discernible surface expressions of harm measured by known marine life impacts may not ultimately prove to be a fair basis for comparison.

Next Steps in the Damage Assessment and Restoration Process

The data collected as part of the damage assessment process will at some point be evaluated by resource specialists for both the trustees and BP. Given that there is no way to exactly quantify the extent of shoreline oiling or the number of birds or other wildlife impacted, the final damage assessment will inevitably consist of estimates developed on the basis of careful examination of the field data (including on-going studies), comparisons to existing baseline data, reviews of the relevant literature, and much debate among the parties involved. Best professional judgment will be needed where data gaps or uncertainty exist.

When the trustees reach a conclusion as to the extent and nature of the damages that occurred and the appropriate amount of restoration required to compensate for the damages, then the matter may proceed to litigation, and be resolved by either court order or settlement. Or, in the spirit of cooperative assessment, the parties may reach a settlement without any litigation. Once a settlement is reached, depending on the terms of the settlement, the responsible party may have two choices. It can opt to implement the required amount of restoration with trustee oversight, or it can pay the trustees to implement the required restoration. Either way, the terms of the agreement are memorialized through a consent decree which must be approved by the Department of Justice.

Finally, with numerous studies ongoing, both under the auspices of the formal damage assessment process and outside it, the published literature regarding environmental impacts from the *Deepwater Horizon* spill can be expected to grow substantially in the months and years ahead.

SOURCES OF ADDITIONAL INFORMATION ON THE NRDA PROCESS

- NOAA's Damage Assessment Remediation and Restoration Program: http://www.gulfspillrestoration.noaa.gov/
- NOAA's Office of Response and Restoration: www.deepwaterhorizon.noaa.gov
- US Fish and Wildlife Service: http://www.fws.gov/home/dhoilspill/restoration
- Online tool that provides you with near-real time information about the response effort. Developed by NOAA with the Environmental Protection Agency, U.S. Coast Guard, and the Department of the Interior, the site offers you a "one-stop shop" for spill response information: www.geoplatform.gov
- Data from the Department of Energy, the Environmental Protection Agency, the National Oceanic and Atmospheric Administration, the Department of the Interior, and the states of Florida and Louisiana related to the spill, its effects, and the cleanup effort: www.data.gov/restorethegulf

End Notes

[1] In the case of offshore facilities, "responsible party" is defined as the "lessee or permittee of the area in which the facility is located or the holder of the right of use and easement granted under applicable State law or the Outer Continental Shelf Lands Act for the area in which the facility is located (if the holder is a different person than the lessee or permittee) . . ."). 33 U.S.C. § 2701(32).

[2] 33 U.S.C. § 2702(b)(2)(A). Trustees act "on behalf of the public" as trustees for natural resources. Federal trustees are designated by the President. State trustees are designated by their Governors. Affected Tribal and foreign nations can also claim trustee authority. 33 U.S.C. § 2706(b).

[3] 33 U.S.C. § 2706 (d)(1).

[4] Ibid. at § 2706(d).

[5] 15 C.F.R. part 990.

[6] 42 U.S.C. §§ 9601 et seq. CERCLA, in turn, built on provisions in the Clean Water Act Amendments of 1977, which first codified federal authority to recover damages for natural resources. Specifically, CERCLA provided additional direction concerning the measure of damages, the use and effect of natural resource damage assessments, and the designation of trustees.

[7] BP is a responsible party, but other companies involved may be named also once liability issues are established. For the time being, BP has entered into a funding agreement with the

Natural Resource Damage Trustees, and is currently reimbursing their costs. If, ultimately, other responsible parties are named, the Trustees must find them jointly and severally liable under the law.

[8] Under CERCLA, by contrast, damage assessments were carried out, for the most part, in a non-cooperative and adversarial manner. Since the trustees could essentially dictate how damages would be determined, responsible parties typically opted to conduct their own assessment in preparation for a court defense should the case end up in litigation. CERCLA regulations (and OPA itself) impose a "rebuttable presumption" in favor of the trustee's damage assessment: If the responsible party disagrees with the trustee's assessment, it bears the burden of proving that the assessment was wrong. This regulatory arrangement often set the stage for parallel and dueling assessments, as emerged in the aftermath of the *Exxon Valdez* incident.

[9] NOAA, *Preassessment Phase: Guidance Document for Natural Resource Damage Assessment Under the Oil Pollution Act of 1990* (August 1996), http://www.darrp.noaa.gov/library/pdf/PPD_COV.PDF.

[10] Ibid.

[11] Confidential Commission staff interviews with state and federal trustees.

[12] With the exception of a few coastal states, damage assessment training, resources, and staff are often gathered onthe-fly when a spill occurs. Expertise is pulled from within state agencies.

[13] The same distinction applies in the case of actions brought by state attorneys general on behalf of state agencies.

[14] There are some exceptions to this general rule. A description of the sources and uses of penalties and fines resulting from oil spills is provided in the Oil Spill Commission's Staff Working Paper 14: "Unlawful Discharges of Oil: Legal Authorities for Civil and Criminal Enforcement and Damage Recovery."

[15] In the Exxon *Valdez* case, of the $900 million recovered from Exxon in a civil settlement, roughly one-fourth ($213.1 million) was used to reimburse the federal government and the State of Alaska for costs incurred in damage assessment and spill response. The remaining $686.9 million was spent in Alaska on efforts to restore resources that were directly harmed by the spill (e.g., sea birds, sea otters, whales and their habitat, etc.). If the State and federal government had brought suit solely under criminal fine or civil penalty authority, only a small portion of the funds recovered from Exxon could have been used to restore resources damaged by the spill.

[16] 42 U.S.C. §§ 4321-4347.

[17] 15 C.F.R. § 990.30 (emphasis added).

[18] Ibid.

[19] There are times when an ecosystem experiences a total loss in services and cannot recover naturally. In this case, the services are considered to be lost in perpetuity and compensatory restoration is calculated accordingly.

[20] Email from Cynthia Dohner, U.S. Fish and Wildlife Service, January 20, 2011; "First project to restore waterfowl habitat since oil spill is underway," *WLBT-TV [Jackson, MS]*, January 21, 2011

[21] Notice of Intent to Conduct Restoration Planning (pursuant to 15 C.F.R. Section 990.44) - DISCHARGE OF OIL FROM THE DEEPWATER HORIZON MOBILE OFFSHORE DRILLING UNIT AND THE SUBSEA MACONDO WELL INTO THE GULF OF MEXICO, APRIL 20, 2010, http://www.darrp.noaa.gov/southeast/deepwater_horizon/pdf/Deepwater_Horizon_Final_NOI.pdf.

[22] Once validated, the data are being made public at www.geoplatform.gov and www.data.gov.

[23] NOAA, Deepwater BP Oil Spill, Natural Resource Damage Assessment: NRDA By the Numbers, January 2011, http://www.gulfspillrestoration.noaa.gov/wp-content/uploads/2011/01/Final-NRDA-by-the-Numbers-Jan-20111.pdf

[24] National Incident Command Joint Analysis Group, Review of Preliminary Data to Examine Oxygen Levels In the Vicinity of MC252#1: May 8 to August 9, 2010 (August 16, 2010).

[25] Richard Camilli et al., "Tracking Hydrocarbon Plume Transport and Biodegradation at Deepwater Horizon," Science 330, no. 6001 (2010): 201–204.

[26] Microbes give off carbon dioxide as part of the biodegradation process which, in turn, depletes oxygen in the water column.

[27] Ibid.

[28] Terry Hazen et al., "Deep-Sea Oil Plume Enriches Indigenous Oil-Degrading Bacteria", Science 330, no. 6001 (2010): 208–211.

[29] David Valentine et al., "Propane Respiration Jump-Starts Microbial Response to a Deep Oil Spill," Science 330, no. 6001 (2010): 204–208.

[30] Arne-R. Diercks et al., "Characterization of Subsurface Polycyclic Aromatic Hydrocarbons at the Deepwater Horizon Site," *Geophysical Research Letters* 37 (2010).

[31] *Exxon Valdez* Oil Spill Trustee Council, *Exxon Valdez Oil Spill Restoration Plan: Update on Injured Resources and Services*, September 1996.

In: Gulf Oil Spill of 2010...
Editors: C. R. Walsh, J. P. Duncan

ISBN: 978-1-61324-729-7
© 2012 Nova Science Publishers, Inc.

Chapter 6

COST OF MAJOR SPILLS MAY IMPACT VIABILITY OF OIL SPILL LIABILITY TRUST FUND[*]

*United States Government Accountability Office
Statement of Susan A. Fleming,
Director Physical Infrastructure*

WHY GAO DID THIS STUDY

On April 20, 2010, an explosion at the mobile offshore drilling unit *Deepwater Horizon* resulted in a massive oil spill in the Gulf of Mexico. The spill's total cost is unknown, but may result in considerable costs to the private sector, as well as federal, state, and local governments. The Oil Pollution Act of 1990 (OPA) set up a system that places the liability— up to specified limits—on the responsible party. The Oil Spill Liability Trust Fund (Fund), administered by the Coast Guard, pays for costs not paid for by the responsible party.

GAO previously reported on the Fund and factors driving the cost of oil spills and is beginning work on the April 2010 spill. This testimony focuses on (1) how oil spills are paid for, (2) the factors that affect major oil spill costs,

---

[*] This is an edited, reformatted and augmented version of the United States Government Accountability Office publication, Statement of Susan A. Fleming, Director Physical Infrastructure, GAO-10-795T, dated June 16, 2010.

and (3) implications of major oil spill costs for the Fund. It is largely based on GAO's 2007 report, for which GAO analyzed oil spill cost data and reviewed documentation on the Fund's balance and vessels' limits of liability. To update the report, GAO obtained information from and interviewed Coast Guard officials.

WHAT GAO RECOMMENDS

In 2007, GAO recommended that the Coast Guard (1) adjust liability limits for inflation and (2) determine whether liability limits should vary by vessel type. The Coast Guard agreed with both recommendations and implemented the former but not the latter recommendation.

WHAT GAO FOUND

OPA places the primary burden of liability for the costs of oil spills on the responsible party in return for financial limitations on that liability. Thus, the responsible party assumes the primary burden of paying for spill costs—which can include both removal costs (cleaning up the spill) and damage claims (restoring the environment and compensating parties that were economically harmed). To pay both the costs above this limit and costs incurred when a responsible party does not pay or cannot be identified, OPA authorized use of the Fund, up to a $1 billion per spill, which is financed primarily from a per-barrel tax on petroleum products. The Fund also may be used to pay for natural resource damage assessments and to monitor the recovery activities of the responsible party, among other things. While the responsible party is largely paying for the current spill's cleanup, Coast Guard officials said that they began using the Fund—which currently has a balance of $1.6 billion—in May 2010 to pay for certain removal activities in the Gulf of Mexico.

Several factors, including location, time of year, and type of oil, affect the cleanup costs of noncatastrophic spills. Although these factors will certainly affect the cost of the Gulf spill—which is unknown at this time—in this spill, additional factors such as the magnitude of the oil spill will impact costs.

These factors can affect the breadth and difficulty of recovery and the extent of damage in the following ways:

- Location. A remote location can increase the cost of a spill because of the additional expense involved in mounting a remote response. A spill that occurs close to shore can also become costly if it involves the use of manual labor to remove oil from sensitive shoreline habitat.
- Time of year. A spill occurring during fishing or tourist season might carry additional economic damage, or a spill occurring during a stormy season might prove more expensive because it is more difficult to clean up than one occurring during a season with generally calmer weather.
- Type of oil. Lighter oils such as gasoline or diesel fuels dissipate and evaporate quickly—requiring minimal cleanup—but are highly toxic and create severe environmental impacts. Heavier oils such as crude oil do not evaporate and, therefore, may require intensive structural and shoreline cleanup.

Since the Fund was authorized in 1990, it has been able to cover costs not covered by responsible parties, but risks and uncertainties exist regarding the Fund's viability. For instance, the Fund is at risk from claims resulting from spills that significantly exceed responsible parties' liability limits. Of the 51 major oil spills GAO reviewed in 2007, the cleanup costs for 10 exceeded the liability limits, resulting in claims of about $252 million. In 2006, Congress increased liability limits, but for certain vessel types, the limits may still be low compared with the historic costs of cleaning up spills from those vessels. The Fund faces other potential risks as well, including ongoing claims from existing spills, claims related to sunken vessels that may begin to leak oil, and the threat of a catastrophic spill—such as the recent Gulf spill.

Mr. Chairman,
Ranking Member McCain,
and Members of the Subcommittee:

I appreciate the opportunity to be here today to discuss the costs of major oil spills and the potential impacts on the Oil Spill Liability Trust Fund (Fund). On April 20, 2010, an explosion from a well site at which the mobile offshore drilling unit (MODU), *Deepwater Horizon*, had been drilling resulted in a spill of national significance in the Gulf of Mexico, which is, to date, only partially contained. Since the explosion occurred, oil has been leaking into the Gulf of Mexico at an estimated rate of between 12,000 and 19,000 barrels per day, according to the National Incident Command's Flow Rate Technical Group, making this one of the largest, if not the largest spill in U.S. waters to date.[1] BP, which leased the *Deepwater Horizon* at the time of the explosion, continues to try to contain the leak. The total cost of cleaning up this massive and potentially unprecedented spill, the untold damage to the environment, as well as the potential impact to the livelihood and the economic status of the region, will be undetermined for some time. However, current estimates suggest that spill cleanup and related damages claims will be in the tens of billions of dollars—well beyond the costs of the *Exxon Valdez*. This spill and future spills all have the potential to result in considerable costs to the private sector, as well as federal, state, and local governments.

The Oil Pollution Act of 1990 (OPA),[2] which was enacted after the *Exxon Valdez* spill in 1989, established a "polluter pays" system that places the primary burden of liability for the costs of spills up to a statutory maximum, on the party responsible. OPA also established the Fund to pay for oil spill costs when the responsible party cannot or does not pay.[3] The Fund is financed primarily from a per-barrel tax on petroleum products either produced in the United States or imported from other countries and administered by the National Pollution Funds Center (NPFC) within the U.S. Coast Guard. While this system is well understood, the total costs involved in responding to oil spills are less clear. Costs paid by the Fund are required to be documented and reported, but the costs paid by the party responsible for the spill are not required to be reported.[4] The resulting lack of information about the total cost of spills, the significant claims made on the Fund to cover the costs beyond the established OPA liability limits borne by the responsible party, and the potential impact of a catastrophic spill of unprecedented costs, have all raised concerns about the Fund's long-term viability.

Mr. Chairman, in response to your request, we are just beginning work related to the April 2010 spill and its implications for the Fund. However, we have done considerable work looking at the cost of major spills in recent years and the factors that contribute to making spills particularly expensive to clean up and mitigate. While our previous work focused on spills from vessels and not offshore facilities, it is likely that many of the same factors that we identified that affect the cost of the oil spills will apply to the current oil spill. Additionally, our previous work identified several potential risks to the Fund and made recommendations to the Commandant of the Coast Guard to address some of the risks.

My remarks today are intended to provide a context for looking at the nation's approach to paying the costs of such spills. Specifically, my testimony focuses on (1) how oil spills are paid for, (2) the factors that affect major oil spill costs, and (3) the implications of major oil spill costs for the Oil Spill Liability Trust Fund.[5] My comments are based primarily on our September 2007 report on oil spill costs, which was issued to the Senate Committee on Commerce, Science, and Transportation, and the House Committee on Transportation and Infrastructure.[6] In our 2007 report, we determined that there were 51 major oil spills— with removal costs and damage claims totaling at least $1 million— that occurred in U.S. waters from 1990 through 2006.[7] Collectively, from public and nonpublic sources, we estimated that responsible parties and the Fund have paid between approximately $860 million and $1.1 billion to clean up these spills and compensate affected parties. Responsible parties paid between about 72 to 78 percent of these costs. The 51 major spills (exceeding $1 million in total costs) we identified, which constituted about 2 percent of the 3,389 vessel spills that occurred from 1990 to 2006, varied greatly from year to year in number and cost and showed no discernible trends in frequency or size.[8]

In preparing our September 2007 report we analyzed oil spill removal cost and claims data from NPFC, the National Oceanic and Atmospheric Administration's (NOAA) Damage Assessment, Remediation, and Restoration Program, and the Department of the Interior's (DOI) Natural Resource Damage Assessment and Restoration Program and U.S. Fish and Wildlife Service. We also analyzed cost data obtained from vessel insurers and through contract with Environmental Research Consulting.[9] We also interviewed NPFC, NOAA, and state officials responsible for oil spill response, as well as industry experts and representatives from key industry associations and a vessel owner. In addition, we reviewed documentation from the NPFC regarding the Fund balance and vessels' limits of liability. Earlier this month, we obtained updated information from and interviewed NPFC officials to update our September 2007 report's findings and to gather information on the recent oil spill in the Gulf of Mexico. In addition, we have just started work on the Oil Spill Liability Trust Fund at the request of the Chairman of this Subcommittee and other congressional members.

Summary

OPA establishes a "polluter pays" system that is intended to act as a deterrent by placing the primary burden of liability[10] for the costs of spills on the party responsible for the spill in return for financial limitations on that liability.[11] Under this system, the responsible party assumes, up to a specified limit that is subject to certain conditions, the burden of paying for spill costs—which can include both removal costs (cleaning up the spill) and damage claims (restoring the environment and payment of compensation to parties that were economically harmed by the spill). Above the specified limit, which varies depending on the type of vessel or facility, the responsible party is no longer financially liable. Responsible parties are liable without limit, however, if the oil discharge is the result of gross negligence or willful misconduct, or a violation of federal operation, safety, or construction regulations. To pay costs above the limit of liability, as well as to pay costs when a responsible party does not pay or cannot be identified, OPA authorized use of the Fund, which is financed primarily from a per-barrel tax on petroleum products either produced in the United States or imported from other countries. Offshore facilities' limit of liability is all removal costs plus $75 million for damage claims.[12] The Fund also may be used to pay for natural resource damage assessments and to monitor the recovery activities of the responsible party, among other things. Coast Guard officials said that they began using the Fund in May 2010 to pay for removal activities in the Gulf of Mexico.

Several factors affect the costs of a noncatastrophic spill, according to industry experts and agency officials and the studies we reviewed—the spill's location, the time of year it occurs, and the type of oil spilled. Additionally, the magnitude of the oil spill will also impact costs of the *Deepwater Horizon* spill. A remote location, for example, can increase the cost of a spill because of the additional expense involved in mounting a remote response. Similarly, a spill that occurs close to shore rather than further out at sea can become more expensive because it may involve the use of manual labor to remove oil from sensitive shoreline habitat. Time also has situation-specific effects, in that a spill that occurs at a particular time of year might involve a much greater cost than a spill occurring in the same place but at a different time of year. For example, a spill occurring during fishing or tourist season might carry additional economic damage, or a spill occurring during a typically stormy season might prove more expensive because it is more difficult to clean up than one occurring during a season with generally calmer weather. The specific type of oil affects costs because the type of oil can affect the amount of cleanup needed and the amount of natural resource damage incurred. Lighter oils such as gasoline or diesel fuels dissipate and evaporate quickly—requiring minimal cleanup—but are highly toxic and create severe environmental impacts. Heavier oils such as crude oil do not evaporate and, therefore, may require intensive structural and shoreline cleanup; and while they are less toxic than light oils, heavy oils can harm waterfowl and fur-bearing mammals through coating and ingestion. Each spill's cost reflects the particular mix of these factors, and no factor is clearly predictive of the outcome. Although the total costs of the Gulf Coast spill will be unknown for some time, many of the same key factors such as location, time of year, oil type, and the magnitude of the oil spilled, will certainly impact the costs of this spill. For example, the spill occurred in the spring in an area of the country—the Gulf Coast—that relies heavily on revenue from tourism and the commercial fishing industry. According to one expert, the loss in revenue from suspended commercial and recreational fishing in the Gulf Coast states is currently estimated at $144 million per year.[13]

Since it was authorized in 1990, the Fund has been able to cover costs that responsible parties have not paid from noncatastrophic spills, but risks and uncertainties exist regarding the Fund's viability. In particular, the Fund is at risk from claims resulting from spills that significantly exceed responsible parties' liability limits. The effect of such spills can be seen among the 51 major oil spills we identified in 2007: 10 of them exceeded the limit of liability, resulting in claims of about $252 million on the Fund. In the Coast Guard and Maritime Transportation Act of 2006, Congress increased these liability limits,[14] but additional attention to the limits appears warranted because the liability limits for certain vessel types may still be disproportionately low compared with their historic spill cost. For example, of the 51 major spills since 1990, 15 resulted from tank barges. The average cost for these 15 tank barge spills was about $23 million— more than double the average liability limit ($10.3 million) for these vessels. In its August 2009 report examining oil spills that exceeded the limits of liability, the Coast Guard had similar findings on the adequacy of some of the current limits and their potential effect on the the Fund. Aside from issues related to limits of liability, the Fund faces other potential drains on its resources, including ongoing claims from existing spills, claims related to already-sunken vessels that may begin to leak oil, and the threat of a catastrophic spill—such as the *Deepwater Horizon*—which could have a significant impact on the Fund's viability.

In our September 2007 report, we recommended that the Commandant of the Coast Guard (1) determine whether and how liability limits should be changed, by vessel type, and make recommendations about these changes to Congress and (2) adjust the limits of liability for vessels every 3 years to reflect changes in inflation, as appropriate. The Department of Homeland Security (DHS), including the Coast Guard, generally agreed with the report's contents and agreed with the recommendations. In July 2009, the Commandant of the Coast Guard implemented our recommendation to adjust limits of liability for vessels every 3 years to reflect changes in inflation,[15] but to date, has not implemented our recommendation to determine whether and how liability limits should be changed by vessel type and make recommendations about these changes to Congress. We continue to believe that adjusting liability limits for particular vessel types, notably tank barges, would ensure that the "polluter pays" principle is carried out in practice.

THE PRIMARY BURDEN OF LIABILITY FOR THE COSTS OF OIL SPILLS IS ON THE RESPONSIBLE PARTY, UP TO SPECIFIED LIMITS

OPA establishes a "polluter pays" system that places the primary burden of liability for the costs of spills on the party responsible for the spill in return for financial limitations on that liability. Under this system, the responsible party assumes, up to a specified limit, the burden of paying for spill costs— which can include both removal costs (cleaning up the spill) and damage claims (restoring the environment and payment of compensation to parties that were economically harmed by the spill). Above the specified limit, the responsible party generally is no longer financially liable. Responsible parties are liable without limit, however, if the oil discharge is the result of gross negligence or willful misconduct, or a violation of federal operation, safety, and construction regulations. OPA's "polluter pays" system is intended to provide a deterrent for responsible parties who could potentially spill oil by requiring that they assume the burden of responding to the spill, restoring natural resources, and compensating those damaged by the spill, up to the specified limit of liability. (See table 1 for the limits of liability for vessels and offshore facilities.)

In general, liability limits under the OPA depend on the kind of vessel or facility from which a spill comes. For an offshore facility, liability is limited to all removal costs plus $75 million. For tank vessels, liability limits are based on the vessel's tonnage and hull type. In both cases, certain circumstances, such as gross negligence, eliminate the caps on liability altogether. According to the Coast Guard, the leaking well in the current spill is an offshore facility. As noted earlier, pursuant to OPA, the liability limit for offshore facilities is all removal costs plus $75 million for damage claims. The Coast Guard also notes that liability for any spill on or above the surface of the water in this case would be between $65 million and $75 million. The range derives from a statutory division of liability for mobile offshore drilling units.[16] For spills on or above the surface of the water, mobile offshore drilling units are treated first as tank vessels up to the limit of liability for tank vessels and then as offshore facilities.[17]

For example, if an offshore facility's limit of liability is $75 million (not counting removal costs, for which there is unlimited liability for offshore facilities) and a spill resulted in $100 million in costs, the responsible party has to pay up to $75 million in damage claims—leaving $25 million in costs beyond the limit of liability.[18] Under OPA, the authorized limit on federal expenditures for a response to a single spill is currently set at $1 billion, and natural resource damage assessments and claims may not exceed $500 million. OPA requires that responsible parties must demonstrate their ability to pay for oil spill response up to statutorily specified limits. Specifically, by statute, with few exceptions, offshore facilities that are used for exploring for, drilling for, producing, or transporting oil from facilities engaged in oil exploration, drilling, or production are required to have a certificate of financial responsibility that demonstrates their ability to pay for oil spill response up to statutorily specified limits. If the responsible party denies a claim or does not settle it within 90 days, a claimant may commence action in court against the responsible party, or present the claim to the NPFC.

OPA also provides that the Fund[19] can be used to pay for oil spill removal costs and damages when those responsible do not pay or cannot be located. This may occur when the source of the spill and, therefore, the responsible party is unknown, or when the responsible party does not have the ability to pay. In other cases, since the cost recovery can take a period of years, the responsible party may become bankrupt or dissolved.

Table 1. Description of Vessels and Offshore Facilities and Current Limits of Liability

Vessels	Description	Limit of liability
Oil tanker	An oil tanker is a ship designed to carry oil in large tanks.	Single hull: Vessels greater than 3,000 gross tons: the greater of $3,200 per gross ton or $23,496,000 million. Vessels less than or equal to 3,000 gross tons: the greater of $3,200 per gross ton or $6,408,000 million. Double hull: Vessels greater than 3,000 gross tons: the greater of $2,000 per gross ton or $17,088,000 million. Vessels less than or equal to 3,000 gross tons: the greater of $2,000 per gross on or $4,272,000 million.
Tank barge	A tank barge is a non-self-propelled vessel that carries liquid, solid, or gaseous cargos in bulk in tanks primarily through rivers and inland waterways.	The greater of $1,000 per gross ton or $854,400.
Cargo ship or freighter	A cargo ship or freighter is a vessel that transports non-oil goods and materials.	
Fishing vessel	A fishing vessel is a ship that is used to catch fish for commercial use.	
Offshore facility	An offshore facility is any facility of any kind located in, on, or under any of the navigable waters of the U.S., and any facility of any kind that is subject to the jurisdiction of the U.S. and is located in, on, or under any other waters, other than a vessel or a public vessel.	All cleanup costs plus $75 million.
Mobile offshore drilling unit (MODU)	A mobile offshore drilling unit is a vessel (other than a self-elevating lift vessel) capable of use as an offshore facility.	For a discharge on or above the surface of the water, a MODU is first treated as a tank vessel up to the limit of liability for tank vessels. For costs above the vessel liability limit, the MODU is treated as an offshore facility.

Source: GAO.

NPFC manages the Fund by disbursing funds for federal cleanup, monitoring the sources and uses of funds, adjudicating claims submitted to the Fund for payment, and pursuing reimbursement from the responsible party for costs and damages paid by the Fund. The Coast Guard is responsible for adjusting vessels' limits of liability for significant increases in inflation and for making recommendations to Congress on whether other adjustments are necessary to help protect the Fund.[20] DOI's Minerals Management Service is responsible for adjusting limits of liability of offshore facilities.

Response to large oil spills is typically a cooperative effort between the public and private sector, and there are numerous players who participate in responding to and paying for oil spills. To manage the response effort, the responsible party, the Coast Guard, EPA, and the pertinent state and local agencies form the unified command, which implements and manages the spill response.[21]

OPA defines the costs for which responsible parties are liable and the costs for which the Fund is made available for compensation in the event that the responsible party does not pay or is not identified.[22] These costs, or "OPA compensable" costs, are of two main types:

- Removal costs: Removal costs are incurred by the federal government or any other entity taking approved action to respond to, contain, and clean up the spill. For example, removal costs include the equipment used in the response—skimmers to pull oil from the water, booms to contain the oil, planes for aerial observation—as well as salaries and travel and lodging costs for responders.
- Damages caused by the oil spill: Damages that can be compensated under OPA cover a wide range of both actual and potential adverse effects from an oil spill, for which a claim may be made to either the responsible party or the Fund. Claims include natural resource damage claims filed by trustees, claims for uncompensated removal costs and third-party damage claims for lost or damaged property and lost profits, among other things.[23]

The Fund has two major components—the Principal Fund and the Emergency Fund. The Principal Fund provides the funds for third-party and natural resource damage claims, limit of liability claims, reimbursement of government agencies' removal costs, and provides for oil spill-related appropriations. A number of agencies—including the Coast Guard, EPA, and DOI—receive an annual appropriation from the Principal Fund to cover administrative, operational, personnel, and enforcement costs. To ensure rapid response to oil spills, OPA created an Emergency Fund that authorizes the President to spend $50 million each year to fund spill response and the initiation of natural resource damage assessments, which provide the basis for determining the natural resource restoration needs that address the public's loss and use of natural resources as a result of a spill.

Emergency funds not used in a fiscal year are carried over to the subsequent fiscal years and remain available until expended. To the extent that $50 million is inadequate, authority under the Maritime Transportation Security Act of 2002 grants authority to advance up to $100 million from the Fund to pay for removal activities. These emergency funds may be used for containing and removing oil from water and shorelines, preventing or minimizing a substantial threat of discharge, and monitoring the removal activities of the responsible party. NPFC officials told us in June 2010 that the emergency fund has received the advanced authority of $100 million for the Federal On-Scene Coordinator to respond to the spill and for federal trustees to initiate natural resource damage assessments along with an additional $50 million that had not been apportioned in 2006. Officials said they began using emergency funds at the beginning of May to pay for removal activities in the Gulf of Mexico.[24]

The Fund is financed primarily from a per-barrel tax on petroleum products either produced in the United States or imported from other countries. The balance of the Fund (including both the Principal and the Emergency Fund) has varied over the years (see fig. 1).[25] The Fund's balance generally declined from 1995 through 2006, and from fiscal year 2003 through 2007, its balance was less than the authorized limit on federal expenditures for the response to a single spill, which is currently set at $1 billion. This was in part because the Fund's main source of revenue—a $0.05 per barrel tax on U.S. produced and imported oil—was not collected for most of the time from 1995 through 2006.[26] However, the Energy Policy Act of 2005 reinstated the barrel tax beginning in April 2006. [27] Subsequently, the Emergency Economic Stabilization Act of 2008 increased the tax rate to $0.08 per barrel through 2016.28 The balance in the Fund as of June 1, 2010, was about $1.6 billion.29 With the barrel tax once again in place, NPFC anticipates that the Fund will be able to cover potential noncatastrophic liabilities.30 In 2007 we reported several risks to the Fund, including the threat of a catastrophic spill. Although the Fund's balance has increased, significant uncertainties remain regarding the impact of a catastrophic spill—such as the Deepwater Horizon —or multiple catastrophic spills on the Fund's viability.

SEVERAL FACTORS, INCLUDING LOCATION, TIME OF YEAR, AND TYPE OF OIL, COMBINE IN UNIQUE WAYS AND AFFECT THE COST OF EACH OIL SPILL

Location, time of year, and type of oil are key factors affecting oil spill costs of noncatastrophic spills, according to industry experts, agency officials, and our analysis of spills. Given the magnitude of the current spill, however, the size of this spill will also be a factor that affects the costs. Officials also identified two other factors that may influence oil spill costs to a lesser extent—the effectiveness of the spill response and the level of public interest in a spill. In ways that are unique to each spill, these factors can affect the breadth and difficulty of the response effort or the extent of damage that requires mitigation.

Location Affects Costs in Different Ways

According to state officials with whom we spoke and industry experts, there are three primary characteristics of location that affect costs:

- Remoteness: For spills that occur in remote areas, spill response can be particularly difficult in terms of mobilizing responders and equipment, and they can complicate the logistics of removing oil from the water—all of which can increase the costs of a spill.
- Proximity to shore: There are also significant costs associated with spills that occur close to shore. Contamination of shoreline areas has a considerable bearing on the costs of spills as such spills can require manual labor to remove oil from the shoreline and sensitive habitats. The extent of damage is also affected by the specific shoreline location.
- Proximity to economic centers: Spills that occur in the proximity of economic centers can cost more when local services are disrupted. For example, a spill near a port can interrupt the flow of goods, necessitating an expeditious response in order to resume business activities, which could increase removal costs.

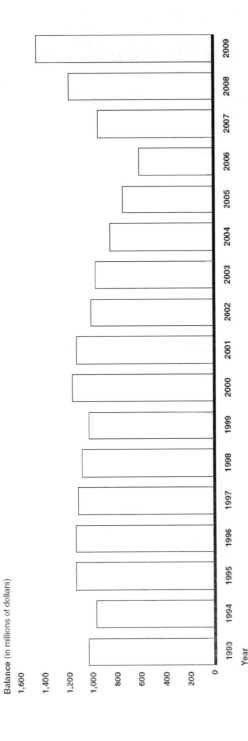

Source: GAO analysis of NPFC data.
Note: The Fund balance increase in 2000 was largely due to a transfer of $181.8 million from the Trans-Alaska Pipeline Liability Fund.

Figure 1. Oil Spill Liability Trust Fund Balance, Fiscal Years 1993-2009.

Time of Year Affects Local Economies and Response Efforts

The time of year in which a spill occurs can also affect spill costs—in particular, affecting local economies and response efforts. According to several state and private-sector officials with whom we spoke, spills that disrupt seasonal events that are critical for local economies can result in considerable expenses. For example, spills in the spring months in areas of the country that rely on revenue from tourism may incur additional removal costs in order to expedite spill cleanup, or because there are stricter standards for clean up, which increase the costs. The time of year in which a spill occurs also affects response efforts because of possible inclement weather conditions such as harsh winter storms and even hurricanes that can result in higher removal costs because of the increased difficulty in mobilizing equipment and personnel to respond to a spill in adverse conditions.

Type of Oil Spilled Affects the Extent of the Response Effort and the Amount of Damage

The different types of oil can be grouped into four categories, each with its own set of effects on spill response and the environment. Lighter oils such as jet fuels, gasoline, and diesel fuel dissipate and evaporate quickly, and as such, often require minimal cleanup. However, these oils are highly toxic and can severely affect the environment if conditions for evaporation are unfavorable. For instance, in 1996, a tank barge that was carrying home-heating oil grounded in the middle of a storm near Point Judith, Rhode Island, spilling approximately 828,000 gallons of heating oil (light oil). Although this oil might dissipate quickly under normal circumstances, heavy wave conditions caused an estimated 80 percent of the release to mix with water, with only about 12 percent evaporating and 10 percent staying on the surface of the water.[31] Natural resource damages alone were estimated at $18 million, due to the death of approximately 9 million lobsters, 27 million clams and crabs, and over 4 million fish.

Heavier oils, such as crude oils and other heavy petroleum products, are less toxic than lighter oils but can also have severe environmental impacts. Medium and heavy oils do not evaporate much, even during favorable weather conditions, and can blanket structures they come in contact with—boats and fishing gear, for example—as well as the shoreline, creating severe environmental impacts to these areas, and harming waterfowl and fur-bearing mammals through coating and ingestion. Additionally, heavy oils can sink, creating prolonged contamination of the sea bed and tar balls that sink to the ocean floor and scatter along beaches. These spills can require intensive shoreline and structural clean up, which is time-consuming and expensive. For example, in 1995, a tanker spilled approximately 38,000 gallons of heavy fuel oil into the Gulf of Mexico when it collided with another tanker as it prepared to lighter its oil to another ship.[32] Less than 1 percent (210 gallons) of the oil was recovered from the sea, and, as a result, recovery efforts on the beaches of Matagorda and South Padre Islands were labor intensive, as hundreds of workers had to manually pick up tar balls with shovels. The total removal costs for the spill were estimated at $7 million.

Other Factors also Affect Spill Costs

In our 2007 report, we also reported that industry experts cited two other factors that also affect the costs incurred during a spill.

- Effectiveness of Spill Response: Some private-sector experts stated that the effectiveness of spill response can affect the cost of cleanup. The longer it takes to assemble and conduct the spill response, the more likely it is that the oil will move with changing tides and currents and affect a greater area, which can increase costs. Some experts said the level of experience of those involved in the incident command is critical to the effectiveness of spill response. For example, they said poor decision making during a spill response could lead to the deployment of unnecessary response equipment, or worse, not enough equipment to respond to a spill. Several experts expressed concern that Coast Guard officials are increasingly inexperienced in handling spill response, in part because the Coast Guard's mission has been increased to include homeland security initiatives.
- Public interest: Several experts with whom we spoke stated that the level of public attention placed on a spill creates pressure on parties to take action and can increase costs. They also noted that the level of public interest can increase the standards of cleanliness expected, which may increase removal costs.

Key Factors Will Likely Influence Cost of Gulf Coast Spill

The total costs of the *Deepwater Horizon* spill in the Gulf of Mexico are currently undetermined and will be unknown for some time even after the spill is fully contained. According to a press release from BP, as of June 7, 2010, the cost of the response amounted to about $1.25 billion, which includes the spill response, containment, relief well drilling, grants to the Gulf states, damage claims paid and federal costs. Of the $1.25 billion, approximately $122 million (as of June 1, 2010) has been paid from the Fund for the response operation, according to NPFC officials.[33] The total costs will not likely be known for a while, as it can take many months or years to determine the full effect of a spill on natural resources and to determine the costs and extent of the natural resource damage. However, the spill has been described as the biggest U.S. offshore platform spill in 40 years, and possibly the most costly.

Our work for this testimony did not include a thorough evaluation of the factors affecting the current spill. However, some of the same key factors that have influenced the cost of 51 major oil spills we reviewed in 2007 will likely have an effect on the costs in the Gulf Coast spill. For example, the spill occurred in the spring in an area of the country—the Gulf Coast—that relies heavily on revenue from tourism and the commercial fishing industry. Spills that occur in proximity of tourist destinations like beaches can result in additional removal costs in order to expedite spill cleanup, or because there are stricter standards for cleanup, which increase the costs. In addition, according to an expert, the loss in revenue from suspended commercial and recreational fishing in the Gulf Coast states is currently estimated at $144 million per year.[34] Another factor affecting spills' costs is the type of oil. The oil that continues to spill into the Gulf of Mexico is a light oil—specifically "light sweet crude" oil—that is toxic and can create long-term contamination of shorelines, and harm waterfowl and fur-bearing mammals. According to the U.S. Fish and Wildlife Service, many species of wildlife face grave risk from the spill, as well as 36 national wildlife refuges that may be affected. In recent testimony, the EPA Deputy Administrator described the *Deepwater Horizon* spill as a "massive and potentially unprecedented environmental disaster."

THE FUND HAS BEEN ABLE TO COVER COSTS NOT PAID BY RESPONSIBLE PARTIES, BUT RISKS AND UNCERTAINTIES REMAIN

To date, the Fund has been able to cover costs from major spills that responsible parties have not paid, but risks and uncertainties remain. We reported in 2007 that the current liability limits for certain vessel types, notably tank barges, may have been disproportionately low relative to costs associated with such spills. In addition, the Fund faced other potential risks to its viability, including ongoing claims from existing spills and the potential for a catastrophic oil spill. The current spill in the Gulf of Mexico could result in a significant strain on the Fund, which currently has a balance of about $1.6 billion.

Further Attention to Limits of Liability Is Needed

The Fund has been able to cover costs from major spills that responsible parties have not paid, but additional focus on limits of liability is warranted. Limits of liability are the amount, under certain circumstances, above which responsible parties are no longer financially liable for spill removal costs and damage claims, in the absence of gross negligence or willful misconduct, or the violation of an applicable federal safety, construction, or operating regulation.[35] If the responsible party's costs exceed the limit of liability, the responsible party can make a claim against the Fund for the amount above the limit. Major oil spills that exceed a vessel's limit of liability are infrequent, but their effect on the Fund can be significant. In our 2007 report, we reported that 10 of the 51 major oil spills that occurred from 1990 through 2006 resulted in limit-of-liability claims on the Fund.[36] These limit-of-liability claims totaled more than $252 million and ranged from less than $1 million to more than $100 million. Limit-ofliability claims will continue to have a pronounced effect on the Fund. NPFC estimates that 74 percent of claims under adjudication that were outstanding as of January 2007 were for spills in which the limit of liability had been exceeded. The amount of these claims under adjudication was $217 million.

In 2007, we identified two key areas in which further attention to these liability limits appeared warranted and made recommendations to the Commandant of the Coast Guard regarding both—the need to adjust limits periodically in the future to account for significant increases in inflation and the appropriateness of some current liability limits. Regarding the need to adjust liability limits to account for increases in inflation, we reported that the Fund was exposed to about $39 million in liability claims for the 51 major spills from 1990 through 2006 that could have been saved if the limits of liability had been adjusted for inflation as required by law, and recommended adjusting limits of liability for vessels every 3 years to reflect significant changes in inflation, as appropriate.[37] Per requirementsin OPA as amended by the Delaware River Protection Act, the Coast Guard published an interim rule in July 2009—made final in January 2010—that adjusted vessels' limits of liability to reflect significant increases in the Consumer Price Index, noting that the inflation adjustments to the limits of liability are required by OPA to preserve the deterrent effect and polluter-pays principle embodied in the OPA liability provisions.[38] DOI has been delegated responsibility by the President to adjust the liability limits for offshore facilities and this responsibility has been redelegated by DOI to the Minerals Management Service.[39] To date, these liability limits have not been adjusted for inflation.

The Coast Guard and Maritime Transportation Act of 2006 significantly increased the limits of liability.[40] Both laws base the liability on a specified amount per gross ton of vessel volume, with different amounts for vessels that transport oil commodities (tankers and tank barges) than for vessels that carry oil as a fuel (such as cargo vessels, fishing vessels, and passenger ships). The 2006 act raised both the per-ton and the required minimum amounts, differentiating between vessels with a double hull, that helps prevent oil spills resulting from collision or grounding, and vessels without a double hull.[41] For example, the liability limit for single-hull vessels larger than 3,000 gross tons was increased from the greater of $1,200 per gross ton or $10 million to the greater of $3,000 per gross ton or $22 million.

However, our analysis of the 51 major spills showed that the average spill cost for some types of vessels, particularly tank barges, was higher than the limit of liability, including the new limits established in 2006.[42] Thus, we recommended that the Commandant of the Coast Guard determine whether and how liability limits should be changed by vessel type, and make specific recommendations about these changes to Congress. In its August 2009 Annual Report to Congress on OPA liability limits, the Coast Guard had similar findings on the adequacy of some of the new limits.[43] The Coast Guard found that 51 spills or substantial threats of a spill have resulted or are likely to result in removal costs and damages that exceed the liability limits amended in 2006. Specifically, the Coast Guard reported that liability limits for tank barges and cargo vessels with substantial fuel oil may not sufficiently account for the historic costs incurred by spills from these vessel types. The Coast Guard concluded that increasing liability limits for tank barges and non tank vessels—cargo, freight, and fishing vessels—over 300 gross tons would increase the Fund balance. With regard to making specific adjustments, the Coast Guard said dividing costs equally between the responsible parties and the Fund was a reasonable standard to apply in determining the adequacy of liability limits.[44] However, the Coast Guard did not recommend explicit changes to achieve either that 50/50 standard or any other division of responsibility.

Other Challenges Could also Affect the Fund's Condition

The Fund also faces several other potential challenges that could affect its financial condition:

- *Additional claims could be made on spills that have already been cleaned up:* Natural resource damage claims can be made on the Fund for years after a spill has been cleaned up. The official natural resource damage assessment conducted by trustees can take years to complete, and once it is completed, claims can be submitted to the NPFC for up to 3 years thereafter.[45]
- *Costs and claims may occur on spills from previously sunken vessels that discharge oil in the future:* Previously sunken vessels that are submerged and in threat of discharging oil represent an ongoing liability to the Fund. There are over 1000 sunken vessels that pose a threat of oil discharge.[46] These potential spills are particularly problematic because in many cases there is no viable responsible party that would be liable for removal costs. Therefore, the full cost burden of oil spilled from these vessels would likely be paid by the Fund.
- *Spills may occur without an identifiable source and, therefore, no responsible party:* Mystery spills also have a sustained effect on the Fund, because costs for spills without an identifiable source—and therefore no responsible party—may be paid out of the Fund. Although mystery spills are a concern, the total cost to the Fund from mystery spills was lower than the costs of known vessel spills in 2001

through 2004. Additionally, none of the 51 major oil spills was the result of discharge from an unknown source.
- *A catastrophic spill could strain the Fund's resources:* In 2007, we reported that since the 1989 *Exxon Valdez* spill, which was the impetus for authorizing the Fund's usage, no oil spill has come close to matching its costs—estimated at $2.2 billion for cleanup costs alone, according to the vessel's owner.[47] However, as of early June, the response for the *Deepwater Horizon* spill had already totaled over $1 billion, according to BP, and to date, the spill has not been fully contained. As a result, the Gulf of Mexico spill could easily eclipse the *Exxon Valdez*, becoming the most costly offshore spill in U.S. history.

The Fund is currently authorized to pay out a maximum of $1 billion on a single spill for response costs, with up to $500 million for natural resource damage claims. Although the Fund has been successful thus far in covering costs that responsible parties did not pay, it may not be sufficient to pay such costs for a spill—such as the *Deepwater Horizon*—that are likely to have catastrophic consequences. While BP has said it will pay all legitimate claims associated with the spill, should the company decide it will not or cannot pay for the costs exceeding their limit of liability, the Fund may have to bear these costs. Given the magnitude of the *Deepwater Horizon* spill, the costs could result in a significant strain on the Fund.

Options for Addressing the Fund's Vulnerabilities

Recently, several options have been identified to address the Fund's vulnerabilities. In particular, the Congressional Research Service (CRS)[48] has identified options to address the vulnerabilities, and Members of Congress have also introduced legislation that would address the risks to the Fund.[49] These options include:

- **Increasing liability limits.** CRS proposes raising the liability caps for vessels so that the responsible party would be required to pay a greater share of the costs before the Fund is used. In addition, S. 3305 proposes raising the liability limit for damage claims related to offshore facilities from $75 million to $10 billion.

- **Increasing the per-barrel tax.** CRS and congressional options include increasing the current per-barrel tax used to generate revenue for the Fund in order to raise the Fund's balance—H.R. 4213 proposes raising the tax from the current $0.08 per barrel to $0.34. According to CRS, this option would increase the likelihood that there is sufficient money available in the Fund if costs exceed the responsible party's liability limits.
- **Including oil owners as liable parties.** CRS suggests expanding the definition of liable parties to include the owner of the oil being transported by a vessel.

In addition, the Administration announced a proposal on May 12, 2010, that addresses several aspects of the response to the *Deepwater Horizon* spill, primarily by changing the way the Fund operates. It includes, among other things, proposals to increase the statutory limitation on expenditures from the Fund for a single oil spill response from $1 billion to $1.5 billion for spill response and from $500 million to $750 million per spill for natural resource damage assessments and claims. In addition, similar to the CRS and congressional proposals, the Administration is proposing an increase on the per-barrel tax to $0.09 this year, 7 years earlier than the current law requires.

Mr. Chairman, this concludes my statement. I would be pleased to respond to any questions you or other Members of the Subcommittee may have.

Acknowledgments

Individuals making key contributions to this testimony include Jeanette Franzel, Heather Halliwell, David Hooper, Hannah Laufe, Stephanie Purcell, Susan Ragland, Amy Rosewarne, Doris Yanger, and Susan Zimmerman.

End Notes

[1] The Flow Rate Technical Group is comprised of federal scientists, independent experts, and representatives from universities around the country. It includes representatives from U. S. Geological Survey, National Oceanic and Atmospheric Administration, Department of Energy, Coast Guard, Department of the Interior's Minerals Management Service, the national labs, National Institute of Standards and Technology, University of California-Berkeley, University of Washington, University of Texas, Purdue University, and several

other academic institutions. BP is not involved in the Flow Rate Technical Group except to supply raw data for the scientists and experts to analyze.

[2] Pub. L. No. 101-380, 104 Stat. 489 (1990).

[3] The Fund also pays for the costs of certain federal agency operations.

[4] The financial activities of the Fund and the resulting fund balance are included in the financial statements and disclosures for the Department of Homeland Security.

[5] The National Oil and Hazardous Substances Pollution Contingency Plan states that any oil discharge that poses a substantial threat to public health or welfare of the United States or the environment or results in significant public concern shall be classified as a major spill. For the purposes of our 2007 report, however, we defined major spills as spills with total removal costs and damage claims that exceed $1 million.

[6] GAO, *Maritime Transportation: Major Oil Spills Occur Infrequently, but Risks to the Federal Oil Spill Fund Remain*, GAO-07-1085 (Washington, D.C.: Sept. 7, 2007). The Coast Guard and Maritime Transportation Act of 2006, Pub. L. No. 109-241, 120 Stat. 516 (2006), directed us to conduct an assessment of the cost of response activities and claims related to oil spills from vessels that have occurred since January 1, 1990, for which the total costs and claims paid was at least $1 million per spill. The mandate required that the report summarize the costs and claims for oil spills that have occurred since January 1, 1990, that total at least $1 million per spill, and the source, if known, of each spill for each year. We were not directed to look at spills from offshore facilities.

[7] Our analysis excluded spills for which final costs were not yet known because all claims had not been addressed.

[8] In order to determine the spill cost estimates for the 51 spills in our 2007 report, we obtained the best available cost data from a variety of sources because private-sector and total costs for cleaning up spills and paying damages are not centrally tracked and maintained. We then combined the information that we collected from these various sources to develop cost estimates for the oil spills. However, because the cost data are somewhat imprecise and the data we collected vary somewhat by source, we presented the cost estimates in ranges. The lower and higher bounds of the range represent the low- and high-end of cost information we obtained. Based on reviews of data documentation, interviews with relevant officials, and tests for reasonableness, we determined that the data were sufficiently reliable for the purposes of our report.

[9] Environmental Research Consulting is a private consulting firm that specializes in data analysis, environmental risk assessment, cost analyses, expert witness research and testimony, and development of comprehensive databases on oil and chemical spills in service to regulatory agencies, nongovernmental organizations, and industry.

[10] In the case of a vessel, the responsible party is "any person owning, operating, or demise chartering the vessel." 33 U.S.C. § 2701(32)(A). In the case of an offshore facility the responsible party is the lessee or permittee of the area in which the facility is located or the holder of a right of use and easement granted under applicable State law or the Outer Continental Shelf Lands Act ... for the area in which the facility is located (if the holder is a different person than the lessee or permittee) ... 33 U.S.C. § 2701(32)(C). NPFC has designated the source of the discharges for this incident as BP Exploration and Production, Inc. as lessee for the area, and Transocean Holdings, Inc., as the owner of the mobile offshore drilling unit, and as such, are responsible parties. To date, only BP is paying costs associated with this spill.

[11] This testimony focuses only on the liability imposed by OPA.

[12] When responsible parties' costs exceed their limit of liability and the limit is upheld— because there was no gross negligence willful misconduct, or violations of federal regulations by the vessel owner or operator—the responsible party is entitled to file a claim on the Fund to be reimbursed for costs in excess of the limit.

[13] McKinney, Larry, *The Deepwater Horizon Oil Spill—Putting a Price on the Priceless*, Harte Research Institute for Gulf of Mexico Studies (Corpus Christi, Tex.: 2010).

[14] 33 U.S.C. § 2704(b). The estimate of $65 million is based on Pub. L. No. 109-241, § 603, 120 Stat. 516, 553 (2006).

[15] 74 Fed. Reg. 31358, July 1, 2009. This interim rule was finalized in January 2010. 75 Fed. Reg. 750, January 6, 2010.

[16] A MODU is a vessel capable of use as an offshore facility.

[17] The estimate of $65 million is based on the tonnage of the *Deepwater Horizon* and thus the liability that would be calculated for it as a tank vessel, and $75 million is the cap on liability for offshore facilities.

[18] When responsible parties' costs exceed their limit of liability and the limit is upheld— because there was no gross negligence or violations of federal regulations by the vessel owner or operator—the responsible party is entitled to file a claim on the Fund to be reimbursed for costs in excess of the limit. The NPFC reviews the claim to determine which costs are entitled to compensation under and the responsible party is reimbursed from the Fund.

[19] The Fund was originally established under the Omnibus Budget Reconciliation Act of 1986, Pub. L. No. 99-509, title VIII, § 8033 (Oct. 21, 1986) (*codified at* 26 U.S.C. § 9509), to fund oil spill response activities, but Congress did not authorize its use until enactment of OPA in 1990.

[20] 33 U.S.C. § 2704(d).

[21] The Incident Command System (ICS) is a standardized response management system that is part of the National Interagency Incident Management System. The ICS is organizationally flexible so that it can expand and contract to accommodate spill responses of various sizes. The ICS typically consists of four sections: operations, planning, logistics, and finance/administration.

[22] 33 U.S.C. § 2702(b). In the case of a vessel, the responsible party is "any person owning, operating, or demise chartering the vessel." 31 U.S.C. § 2701(32)(A). In the case of an offshore facility the responsible party "is the lessee or permittee of the area in which the facility is located or the holder of a right of use and easement granted under applicable State law or the Outer Continental Shelf Lands Act ... for the area in which the facility is located (if the holder is a different person than the lessee or permittee) ...." 31 U.S.C. § 2701 (32)(C).

[23] OPA authorizes the United States, states, and Indian Tribes to act on behalf of the public as natural resource trustees for natural resources under their respective trusteeship. Trustees often have information and technical expertise about the biological effects of pollution, as well as the location of sensitive species and habitats that can assist the federal on-scene coordinator in characterizing the nature and extent of site-related contamination and impacts. Federal Trustees include Commerce, DOI, the Departments of Agriculture, Defense, and Energy, and other agencies authorized to manage or protect natural resources.

[24] Under 33 U.S.C. § 2702, the responsible party is liable for the removal costs and damages that result from an oil spill and thus will be responsible for reimbursing the Fund for these expenses.

[25] OPA consolidated the liability and compensation provisions of four prior federal oil pollution initiatives and their respective trust funds into the Oil Spill Liability Trust Fund and

authorized the collection of revenue and the use of the money, with certain limitations, with regards to expenditures. The prior federal laws regarding oil pollution included the Federal Water Pollution Control Act, the Deepwater Port Act of 1974, the Trans-Alaska Pipeline Authorization Act, and the Outer Continental Shelf Lands Act Amendments of 1978. Congress created the Fund in 1986 but did not authorize collection of revenue or use of the money until it passed OPA in 1990.

[26] The tax expired in December 1994. Besides the barrel tax, the Fund also receives revenue in the form of interest on the Fund's principal revenues from amounts recovered from responsible parties for damages resulting from oil spills, from penalties paid pursuant to the Federal Water Pollution Control Act, the Deepwater Port Act of 1974, or the Trans-Alaska Pipeline Authorization Act, and from certain other sources.

[27] Pub. L. No. 109-58, §1361, 119 Stat. 594 (2005).

[28] Pub. L. No. 110-343, § 405, 122 Stat. 3765, 3860. In 2017, the per-barrel tax increases to $0.09. The tax is scheduled to terminate at the end of 2017.

[29] In 2007, we reported that the balance of the Fund was about $600 million at the end of fiscal year 2006, which at the time, was well below its peak of $1.2 billion in 2000. The decline in the Fund's balance primarily reflected an expiration of the barrel tax on petroleum in 1994. However, the tax was reinstated in 2005 and increased to $0.08 per-barrel in 2008; as a result, the Fund is now at its highest balance.

[30] Related GAO products include GAO, *U.S. Coast Guard National Pollution Funds Center: Improvements Are Needed in Internal Control Over Disbursements*, GAO-04-340R (Washington, D.C.: Jan. 13, 2004); and GAO, *U.S. Coast Guard National Pollution Funds Center: Claims Payment Process Was Functioning Effectively, but Additional Controls Are Needed to Reduce the Risk of Improper Payments*, GAO-04-114R (Washington, D.C.: Oct. 3, 2003).

[31] National Research Council of the National Academies, *Oil in the Sea III: Inputs, Fates, and Effects* (Washington, D.C.: 2003). Numbers do not add to 100 percent due to rounding.

[32] Lightering is the process of transferring oil at sea from a very large or ultra-large carrier to smaller tankers that are capable of entering the port.

[33] Of the $122 million, $4.2 million has been used to by the federal trustees to initiate natural resource damage assessments. Under 33 U.S.C. § 2702, the responsible party is liable for the removal costs and damages that result from an oil spill and thus will be responsible for reimbursing the Fund for these expenses.

[34] McKinney, Larry, *The Deepwater Horizon Oil Spill—Putting a Price on the Priceless*, Harte Research Institute for Gulf of Mexico Studies (Corpus Christi, Tex.: 2010). [35] See 33 U.S.C. § 2704 for a more complete discussion of the liability limits and exceptions.

[36] Additional spills had costs in excess of the vessel's limit of liability, but either the limit was not upheld or no claim was filed by the responsible party.

[37] OPA requires the President, who has delegated responsibility to the Coast Guard, through the Secretary of Homeland Security, to issue regulations not less often than every 3 years to adjust the limits of liability to reflect significant increases in the Consumer Price Index. Congress reiterated this requirement in the Coast Guard and Maritime Transportation Act of 2006 by requiring that regulations be issued 3 years after the enactment of the act (July 11, 2006) and every 3 years afterward to adjust the limits of liability to reflect significant increases in the Consumer Price Index.

[38] 74 Fed. Reg. 31358, July 1, 2009.

[39] Executive Order 12777, October 18, 1991, and *Department of the Interior Organization Manual*, Part 118, Chapter 1, Section 1.2, June 18, 2008.

[40] Pub. L. No. 109-241, § 603, 120 Stat. 516, 554. Vessels' liability limits were raised again in 2009 by the Coast Guard to reflect significant increases in inflation, as required by OPA. However, the 2006 adjustment in liability limits, which increased an average of 125 percent for the 51 vessels involved in major oil spills, were substantially higher than the rise in inflation during the period.

[41] OPA requires that all tank vessels (greater than 5,000 gross tons) constructed (or that undergo major conversions) under contracts awarded after June 30, 1990, operating in U.S. navigable waters must have double hulls. Of the 51 major oil spills, all 24 major spills from tank vessels (tankers and tank barges) involved single-hull vessels.

[42] The 15 tank barge spills and the 12 fishing/other vessel spills in our review had average costs greater than both the 1990 and 2006 limits of liability. For example, for tank barges, the average cost of $23 million was higher than the average limit of liability of $4.1 million under the 1990 limits and $10.3 million under the new 2006 limits.

[43] U.S. Coast Guard, *Oil Pollution Act Liability Limits: Annual Report to Congress, Fiscal Year 2009* (Aug. 18, 2009).

[44] We did not assess the reasonableness of adopting such a standard in determining liability limits

[45] 33 U.S.C. § 2712((h)(2). Federal response costs for spills that resulted from hurricanes Katrina and Rita were paid from the Stafford Act Disaster Relief Funds. However, private parties can seek reimbursement from the Fund for cleanup costs and damages in the future. According to NPFC, as of June 2010, claims related to Katrina and Rita have been relatively minor.

[46] Michel, J., D. Etkin, T. Gilbert, J. Waldron, C. Blocksidge, and R. Urban; 2005. *Potentially Polluting Wrecks in Marine Waters: An Issue Paper Prepared for the 2005 International Oil Spill Conference.*

[47] The *ExxonValdez* only discharged about 20 percent of the oil it was carrying. A catastrophic spill from a vessel could result in costs that exceed those of the *Exxon Valdez*, particularly if the entire contents of a tanker were released in a 'worst-case discharge' scenario.

[48] Congressional Research Service, *Oil Spills in U.S. Coastal Waters: Background, Governance, and Issues for Congress* (Washington, D.C.: 2010).

[49] S. 3305, S. 3306, and H.R. 4213, 111th Cong. 2010.

In: Gulf Oil Spill of 2010...
Editors: C. R. Walsh, J. P. Duncan

ISBN: 978-1-61324-729-7
© 2012 Nova Science Publishers, Inc.

Chapter 7

UNLAWFUL DISCHARGES OF OIL: LEGAL AUTHORITIES FOR CIVIL AND CRIMINAL ENFORCEMENT AND DAMAGE RECOVERY[*]

National Commission on the BP Deepwater Horizon Oil Spill and Offshore Drilling

STAFF WORKING PAPER NO. 14

Staff working papers are written by the staff of the National Commission on the BP Deepwater Horizon Oil Spill and Offshore Drilling for the use of members of the Commission. They do not necessarily reflect the views of the Commission or any of its members. In addition, they may be based in part on confidential interviews with government and non-government personnel.

I. INTRODUCTION

The purpose of this staff working paper is to provide an overview of the sources and uses of penalties and other funds recovered as a result of

---

[*] This is an edited, reformatted and augmented version of a National Commission on the BP Deepwater Horizon Oil Spill and Offshore Drilling Publication, Staff Working Paper No. 14.

unpermitted discharges of oil.[1] There are a number of provisions in federal environmental statutes that authorize the federal and/or state governments to seek fines and penalties for violations, recover clean up and removal costs, and secure funds to restore natural resources.

Relevant federal statutes include the Clean Water Act, Oil Pollution Act of 1990, Endangered Species Act, Marine Mammal Protection Act, and Migratory Bird Treaty Act. This paper is intended to cover the primary federal enforcement and recovery authorities, but is not comprehensive. For example, it does not address criminal acts not covered under environmental and wildlife statutes, or state laws applicable to oil spills.

In June 2010, the Department of Justice (DOJ) announced that the United States would pursue all available civil remedies and consider appropriate criminal penalties to ensure that the responsible parties are held accountable for the Deepwater Horizon oil spill.[2] On December 15, 2010, the United States filed a civil complaint in the Eastern District of Louisiana naming, among others, BP, Anadarko, MOEX Offshore 2007, and Transocean as defendants. The complaint seeks: (1) to impose civil penalties on the defendants for alleged violations of the Clean Water Act and (2) "a declaration that the [d]efendants are responsible and strictly liable for unlimited removal costs and damages under the Oil Pollution Act of 1990." The United States expressly reserved its right to amend the complaint "by, among other things, adding new claims and new defendants."[3]

This staff working paper covers the following topics:

1) Clean Water Act civil and criminal enforcement mechanisms, including Supplemental Environmental Projects;
2) Sources and uses of the Oil Spill Liability Trust Fund established by the Oil Pollution Act;
3) Authority to recover removal, clean up, and natural resource damages costs;
4) Limitations on a responsible party's liability under the Oil Pollution Act; and
5) Authority for civil and criminal actions under federal wildlife statutes.

This paper's conclusion lists sources of funds that may be directed to support Gulf of Mexico restoration efforts.

II. CLEAN WATER ACT

The Clean Water Act authorizes the Environmental Protection Agency (EPA) and Coast Guard to assess administrative penalties. It also authorizes DOJ, on behalf of EPA and Coast Guard, to bring lawsuits seeking civil and criminal penalties.[4]

A. Civil Administrative Penalties

Section 311(b)(6) of the Clean Water Act authorizes EPA and the Coast Guard to assess administrative penalties (either Class I or Class II, depending on level of severity) for unpermitted discharges of oil. A Class I civil penalty may not exceed $16,000 per violation, up to a maximum total penalty of $37,500. A Class II civil penalty may not exceed $16,000 "per day for each day during which the violation continues[,]" up to a maximum of $177,500.[5] In the case of the Deepwater Horizon spill, an administrative action to assess a Class II penalty, in lieu of a civil judicial action, would likely yield a substantially smaller penalty, because it would preclude the government from later filing suit pursuant to Section 311(b)(7) to recover a penalty for each barrel of oil released.[6]

B. Civil Judicial Actions

1. Penalties

Section 311(b)(7) of the Clean Water Act provides for civil penalties for unpermitted discharges of oil of up to $37,500 per day of violation or up to $1,100 per barrel of oil discharged.[7] If the unlawful discharge is the result of gross negligence or willful misconduct of the owner, operator, or any person in charge of a vessel or offshore facility, the penalty is not less than $140,000, and not more than $4,300 per barrel of oil discharged.[8] EPA has interpreted these provisions to mean that the government may elect whether per day or volumetric penalties may apply according to how it pleads its case, or plead both approaches in the alternative. The following factors must be considered when courts determine the amount of a civil judicial penalty (or when EPA or Coast Guard determines the amount of an administrative penalty):

the seriousness of the violation or violations, the economic benefit to the violator, if any, resulting from the violation, the degree of culpability involved, any other penalty for the same incident, any history of prior violations, the nature, extent, and degree of success of any efforts of the violator to minimize or mitigate the effects of the discharge, the economic impact of the penalty on the violator, and any other matters as justice may require.[10]

Given the severity of the Deepwater Horizon spill, the civil judicial penalties may be extremely high. News sources have estimated that the maximum civil penalty could be between $4.5 billion and $21 billion.[11] Penalties collected by the United States for civil violations of Section 311 of the Clean Water Act must be deposited into the Oil Spill Liability Trust Fund (described in Part III below).[12]

2. Supplemental Environmental Projects

A civil judicial penalty may be reduced if the defendant agrees to fund or perform a Supplemental Environmental Project (SEP) or similar initiative. SEPs are defined as "environmentally beneficial projects which a defendant/respondent agrees to undertake in settlement of an enforcement action, but which the defendant/respondent is not otherwise legally required to perform."[13] EPA includes SEPs in environmental enforcement settlements to induce defendants to voluntarily improve environmental quality after violations occur.[14]

According to EPA, "the primary purpose of [SEPs] is to encourage and obtain environmental and public health protection and improvements that may not otherwise have occurred without the settlement incentives provided by this [p]olicy."[15] EPA's SEP policy encourages use of SEPs in communities where environmental justice may be an issue. Thus, SEPs offer an opportunity for environmental improvement beyond compliance and are intended to benefit communities where violations occur.[16]

SEPs must be consistent with the provisions of the underlying environmental statutes, advance the objectives of the statutes that are the basis of the enforcement action, have a nexus to the violations, and be explained in detail in an environmental enforcement settlement.[17] EPA may not manage or control SEP funds or "retain authority to manage or administer the SEP."[18]

In addition, federal agencies may not authorize SEPs for statutory duties or particular activities for which Congress has otherwise made funds available.[19] Finally, a defendant may not use a SEP to comply with obligations otherwise required by law.[20]

If a defendant agrees to conduct a SEP, EPA may reduce the civil penalty that the defendant would otherwise be obligated to pay. EPA uses a combination of quantitative and qualitative factors to determine the mitigation percentage and final settlement amount.[21]

C. Criminal Penalties (Fines and/or Imprisonment)

1. Criminal Penalty Authority
Section 309(c) of the Clean Water Act authorizes criminal prosecution for unpermitted oil discharges. Those subject to criminal prosecution include individuals, corporations, associations, states, municipalities, and responsible corporate officers.[22] Depending on the level of intent, the following types of violators may be fined, imprisoned, or both:

- A person who *negligently* violates the oil discharge provisions of Clean Water Act Section 311 for the first time may be punished by a fine of between $2,500 and $25,000 per day of violation, imprisonment for up to a year, or both. For subsequent convictions, penalties increase up to a maximum of $50,000 per day of violation, two years of imprisonment, or both.[23]
- A person who *knowingly* violates Section 311 for the first time is punishable by a fine ranging from $5,000 to $50,000 per day of violation, imprisonment for up to three years, or both. For subsequent convictions, the maximum penalty is $100,000 per day of violation, imprisonment for six years, or both.[24]
- A person who *knowingly* violates Section 311 and who "knows at the time that he thereby places another person in imminent danger of death or serious bodily injury" (i.e. knowing endangerment) shall be subject to a fine of not more than $250,000, imprisonment for up to 15 years, or both.[25] If the "person" is an organization, it shall be subject to a fine of not more than $1,000,000. After a first conviction, the maximum punishment for a subsequent conviction is "doubled with respect to both fine and imprisonment."[26]

Fines collected by the United States for criminal violations of Section 311 of the Clean Water Act must be paid into the Oil Spill Liability Trust Fund (described in Part III below).[27]

2. Criminal Sentences and Fines: Other Factors

When imposing a criminal sentence, courts must consider factors that include the nature and circumstances of the offense; the history and characteristics of the defendant; the need for the sentence imposed to provide just punishment, promote respect for the law, assure adequate deterrence, and protect the public from future crimes of the defendant; and the need to provide restitution to any victims of the offense.[28]

In assessing a criminal fine, courts must consider, among other things, the defendant's income, earning capacity, and financial resources; the burden a fine will impose on the defendant; any monetary loss inflicted on others as a result of the offense; whether restitution is ordered and the amount of such restitution; whether the defendant can pass on the expense of the fine to consumers or other persons; and, if the defendant is an organization, the size of the organization and any measures taken to discipline any agent responsible for the offense.[29]

Courts may order restitution in any criminal case where required or allowed by law and to the extent agreed to by the parties in a plea agreement.[30] Restitution may be paid to the government for use in restoring natural resources.[31] Courts may not impose a fine or penalty to the extent it will impair the defendant's ability to pay restitution.[32]

The Alternative Fines Act authorizes criminal fines for organizations up to the greatest of: (a) the amount specified by statute; (b) $200,000 for a misdemeanor not resulting in death and $500,000 for a felony; or (c) twice the gross gain or gross loss resulting from the offense.[33] If applied in the case of the Gulf spill, the final provision could allow for the imposition of immense criminal fines equaling twice the aggregate losses resulting from the spill.

U.S. Sentencing Guidelines allow courts to impose a community service payment as an additional component of a sentence in order to repair harms caused by a defendant's actions.[34] Such payments punish the defendant, deter similar future conduct, and benefit the public because the payment goes to remedy the impacts of the violation. Community service payments occur with some regularity in criminal environmental cases.[35]

D. Exxon-Valdez Criminal Plea Agreement

In 1991, Exxon pled guilty to violations of the Clean Water Act, Migratory Bird Treaty Act, and Refuse Act, and agreed to a $150 million criminal fine (later reduced to $25 million in recognition of Exxon's cooperation in cleaning up the spill). Of the $25 million, $13 million went to the Crime Victims Fund and $12 million went to the North American Wetlands Conservation Fund. Exxon also paid $100 million in criminal restitution for restoration, which was split evenly between the federal government and Alaska.[36]

III. OIL SPILL LIABILITY TRUST FUND

Congress established the Oil Spill Liability Trust Fund (Trust Fund or Fund) as a funding source for removal costs and damages resulting from oil discharges.[37] The Fund is located within the U.S. Treasury and is managed by the National Pollution Funds Center, an independent unit within the Coast Guard.

A. Funding Sources

The Trust Fund receives funding from the following sources: transfers from pre-existing funds; a per barrel tax on petroleum produced in or imported to the United States; cost recovery from responsible parties; civil and criminal penalties collected under Section 311 of the Clean Water Act; penalties for violations of certain other statutes; and investment interest on the Fund's principal.[38]

In October 2008, Congress raised the per barrel tax from $0.05 to $0.08 until January 1, 2017, and to $0.09 from January 1, 2017 until December 31, 2017.[39] It also removed a provision that phased out the per barrel tax if the balance of the Fund reached $2.7 billion.[40] With this provision removed, the balance of the Fund has the potential to be much larger. The per barrel tax is the largest source of income.[41]

The Fund has two components: (1) the Emergency Fund, which contains $50 million that may be used for removal activities and initiation of natural resource damage assessments without further appropriation; and (2) the Principal Fund, which funds the remaining expenditures.[42] Congress amended the Oil Pollution Act in June 2010 to allow the Coast Guard to obtain one or more advances from the Fund of up to $100 million each to cover Deepwater Horizon removal costs.[43]

B. Trust Fund Expenditures

Uses of the Trust Fund that are relevant here include:

- payment of removal costs incurred by federal and state governments;
- payment of costs of natural resource damage assessments as well as development and implementation of restoration plans;
- payment of claims by individual persons or governments for removal costs and damages;
- payment of costs of federal agencies to administer and enforce the Oil Pollution Act; and
- research and development.[44]

Money from the Trust Fund is also appropriated by Congress to the Denali Commission (to repair or replace storage tanks in Alaska) and to the Prince William Sound Oil Spill Recovery Institute.

Sums expended by the Fund for response, removal, and recovery may be recovered from the responsible party. Recovered response and removal costs are not used to respond to the incident for which they were collected; rather, recovered funds go to the Principal Fund for use in future spills. When no responsible party is identified, the Trust Fund finances response, clean up, and claims. Expenditures are limited to $1 billion per oil pollution incident. Natural resource damage claims and assessments are limited to $500 million per incident.[45]

IV. REMOVAL COSTS AND NATURAL RESOURCES DAMAGES AUTHORITY

In addition to civil and criminal liability under the Clean Water Act, responsible parties are liable for two types of costs under the Oil Pollution Act of 1990: *removal costs* including (a) all removal costs incurred by the United States, a state, or an Indian tribe under provisions of the Clean Water Act, the Intervention on the High Seas Act, or state law; and (b) removal costs incurred by any person which are consistent with the National Contingency Plan;[46] and *damages* for injury to natural resources, injury to real or personal property, loss of subsistence use of natural resources, loss of revenues, profits, and earning capacity due to the destruction of real or personal property or natural resources, and costs of increased public services during or after removal.[47]

Authority to recover damages for injuries to natural resources resulting from the spill is addressed in a separate staff working paper entitled Natural Resource Damage Assessment: Evolution, Current Practice, and Preliminary Findings Related to the Deepwater Horizon Oil Spill (No. 17). In brief, the Oil Pollution Act authorizes natural resource trustees—representing federal, state, Indian, or foreign governments—to recover "the cost of restoring, rehabilitating, replacing, or acquiring the equivalent of, the damaged natural resources; the diminution in value of those natural resources pending restoration; plus the reasonable cost of assessing those damages."[48] Natural resource damages are deposited in a revolving trust account "without further appropriation, for use only to reimburse or pay costs incurred by the trustee . . . with respect to the damaged natural resources."[49]

V. LIMITATION OF LIABILITY UNDER THE OIL POLLUTION ACT OF 1990

Responsible parties are not liable for the costs of removal or damages if violations are caused solely by an act of God, act of war, or act or omission of a third party.[50] Based on the Deepwater Horizon's status as a mobile offshore drilling unit, gross tonnage, and current response costs, the Oil Pollution Act limits the responsible party's liability to all removal costs plus $75 million.[51] The limit would not apply if the incident was proximately caused by a responsible party's gross negligence, willful misconduct, or violation of applicable Federal safety, construction, or operation regulation.[52]

BP "has chosen to waive the statutory limitation on liability under [the Oil Pollution Act]".[53] As of December 22, 2010, BP and the Gulf Coast Claims Facility (administrator of a $20 billion escrow account funded by BP) had paid or approved for payment over $4.3 billion for private claims, removal costs, and direct payments to governments.[54] Congress, the Administration and private citizens have questioned whether current Oil Pollution Act limits on liability are appropriate and whether they should be raised or removed entirely.

Even if BP had not waived the limitation on liability, however, it would still be responsible for paying amounts far greater than $75 million because the limitation on liability also does not apply to civil and criminal penalties under federal and state law, oil spill removal costs under federal law, or claims for damages brought under state law.[55] For BP, these amounts outside the scope of the liability limitation include: removal costs (BP has paid over $1 billion to date);[56] the potential billions of dollars in civil and criminal penalties identified in Section II, above; unlimited liability for damages in some states and potentially, civil and/or criminal penalties under state law. Thus, in the case of the BP spill and other spills, the liability cap would not represent the total amount a responsible party must in fact pay in connection with a spill.

VI. SELECTED WILDLIFE STATUTES

The Endangered Species Act, the Marine Mammal Protection Act, and the Migratory Bird Treaty Act authorize the government to bring both civil and criminal suits for actions that harm protected wildlife. Below is a brief description of each of these laws, which are potentially applicable to the Gulf spill.[57]

A. Endangered Species Act

Under the Endangered Species Act, it is illegal to "take" a species listed as endangered or threatened.[58] Persons who *knowingly* violate civil provisions of the Act may be subject to penalties ranging from $13,200 to $32,500 per offense.[59] Persons who *knowingly* violate criminal provisions of the Act may be fined up to $50,000 and imprisoned for up to one year.[60] Anyone who otherwise violates the Act is subject to a civil penalty up to $650 per offense.[61]

Penalties and fines collected for violations of the Act may be used to pay costs up to $500,000 incurred by a person providing temporary care for fish, wildlife, or plants pending the outcome of a legal proceeding under the Act.[62] Excess funds are deposited in the Cooperative Endangered Species Conservation Fund, which provides grants to states and territories for species and habitat conservation.[63]

B. Marine Mammal Protection Act

Subject to exceptions, the Marine Mammal Protection Act makes it unlawful to take marine mammals.[64] Any person who *knowingly* violates the Act is subject to criminal prosecution and may be fined up to $20,000, imprisoned for up to a year, or both for each violation.[65] Under the civil enforcement provisions, violations are punishable by a penalty of up to $11,000 per violation.[66] Civil penalties under the Act could be substantial for this spill because of the broad statutory definitions of "take" and "harassment."[67]

Criminal fines for violations of the Act may be used by the U.S. Fish and Wildlife Service for protection and recovery of manatees, polar bears, sea otters, and walruses.[68] Civil penalties under the Act are paid to the U.S. Treasury.

C. Migratory Bird Treaty Act

The Migratory Bird Treaty Act makes it a criminal act "to pursue, hunt, take, capture, kill, attempt to take, capture, or kill, possess . . . any migratory bird, [or] any part, nest, or eggs of any such bird . . . ."[69] Under the Act's misdemeanor provisions, violators are strictly liable and may be fined up to $15,000 and imprisoned for up to six months.[70] The Act also permits felony prosecutions,[71] but a person cannot be convicted of a felony under the Act without intent to take a migratory bird, to sell or barter a taken migratory bird, or to place bait in order to take a migratory bird.

Criminal penalties under the Act have historically varied in severity.[72] Although Exxon paid a fine following the Exxon Valdez spill in part based on violations of the Migratory Bird Treaty Act, it is not clear what portion of the fine was imposed pursuant to the Act because Exxon's plea involved a number of violations of federal law.[73] The Act may be an effective tool for assessing criminal fines resulting from the spill.

Criminal fines are generally paid to the Crime Victims Fund.[74] Congress is authorized to direct fines and penalties received for violations of the Act to the North American Wetlands Conservation Fund for wetlands projects that benefit migratory birds and other wildlife.[75]

VII. CONCLUSION

Some applicable federal legal authorities provide that recovered funds may be directed to the Gulf. Others may direct funds to unrelated environmental uses. Below is a list of how funds recovered under specific authorities may be used.

- Reimbursement of clean up and removal costs under the Oil Pollution Act are directed to the Oil Spill Liability Trust Fund;
- Natural Resource Damage recoveries obtained under the Oil Pollution Act are directed to the resources in the Gulf that were injured as a result of the spill;
- Civil penalties under the Clean Water Act are deposited into the Oil Spill Liability Trust Fund but are not specifically designated for restoration or other activities in the Gulf;
- A Clean Water Act civil settlement could include a Supplemental Environmental Project, which would likely aim to restore and enhance the Gulf and coastal areas;
- Criminal fines under the Clean Water Act are deposited into the Oil Spill Liability Trust Fund. A court could also order restitution and/or community service payments, both of which could be used in the Gulf;
- Criminal fines under the Migratory Bird Treaty Act may be directed to the North American Wetlands Conservation Fund, which provides funds for wetlands conservation projects;
- Fines and penalties under the Endangered Species Act may go to the Cooperative Endangered Species Conservation Fund for species and habitat conservation; and
- Criminal fines under the Marine Mammal Protection Act may be used by the U.S. Fish and Wildlife Service for protection and recovery of manatees, polar bears, sea otters, and walruses.

End Notes

[1] Consistent with the Executive Order establishing the Commission, this staff working paper does not attempt to assign or apportion legal responsibility or liability among any of the parties involved in the blowout, explosion, fire, or oil spill.

[2] Press Conference, United States Attorney General Eric Holder, New Orleans, LA (June 1, 2010), http://www.justice.gov/ag/speeches/2010/ag-speech

[3] Complaint of the United States of America, United States v. BP Exploration & Production Inc., No. 10-cv-4536 (E.D. La., Dec. 15, 2010).

[4] In addition to the federal government's authority to bring enforcement actions, the Clean Water Act has a citizen suit provision. Specifically, the Clean Water Act provides for citizen enforcement of violations of "an effluent standard or limitation" or "an order issued by the Administrator or a State with respect to such a standard or limitation . . . ." 33 U.S.C. § 1365(a)(1). Under the Act, a "citizen" is an individual, corporation, partnership, association, state, municipality, commission, political subdivision of a state, or interstate body "having

an interest which is or may be adversely affected." *Id.* § 1365(g). Citizens do not have authority to bring suit under Section 311(unpermitted oil discharges) but may bring suit under Section 301 (general unpermitted discharges). *Id.* § 1356(f). Daily penalties are the same under both Sections, but Section 311 has additional penalty provisions based upon the number of barrels of oil released. *See* 33 U.S.C. § 1321(b)(7). At least 60 days before citizens can file suit, notice must be given to EPA, the state in which alleged violations occur, and the alleged violator. A citizen may not file suit if, after the 60 day notice period, EPA or the state at issue has already commenced a similar action. 33 U.S.C. § 1365(b). EPA may intervene in any citizen suit as a matter of right. *Id.* § 1365(c)(2). If the United States is not party to a citizen suit, it must be given 45 days to review any consent judgment before it may be entered. *Id.* at 1365(c)(3). Citizens may not sue for wholly past violations. To date, several citizen suits have been filed as a result of the Deepwater Horizon oil spill.

[5] 33 U.S.C. § 1321(b)(6)(B); 40 C.F.R. § 19.4.
[6] 33 U.S.C. § 1321(b)(6)(E).
[7] 33 U.S.C. § 1321(b)(7)(A); 40 C.F.R. § 19.4.
[8] 33 U.S.C. § 1321(b)(7)(D); 40 C.F.R. § 19.4.
[9] EPA Office of Enforcement and Compliance Assurance, CIVIL PENALTY POLICY FOR SECTION 311(b)(3) AND SECTION 311(j) OF THE CLEAN WATER ACT 2 (Aug. 1998), http://epa.gov/compliance
[10] 33 U.S.C. § 1321(b)(8).
[11] Joel Achenbach and David Fahrenthold, *Oil spill dumped 4.9 million barrels into Gulf of Mexico, latest measure shows*, WASH. POST (Aug. 3, 2010) (civil penalty range is between $4.5 billion and $18 billion, based on an estimated 4.1 million barrels discharged into Gulf waters); Jonathan Tilove, *BP disputes government estimates of volume of Gulf of Mexico oil spill*, TIMES-PICAYUNE (Dec. 3, 2010) (civil penalties could be as high as $21 billion, based on estimate of 4.9 million barrels of oil discharged from the Macondo well). These ranges are based on the civil penalty per barrel amount (for both negligent and grossly negligent discharges) multiplied by the estimated total number of barrels of oil released. According to the current official government estimate, the Macondo well released approximately 4.9 million barrels of oil over the course of the spill (±10 percent), over 800,000 barrels of which were captured at the wellhead using the top hat and other devices. Deepwater Horizon MC252 Gulf Incident Oil Budget (Aug. 1, 2010), http://www.noaanews.noaa.gov/stories2010/PDFs/DeepwaterHorizonOil-Budget20100801.pdf). BP has not released its own estimate for the total amount of oil discharged from the Macondo well, but it disputes the government's figures on the grounds that, among other things, they fail to take into account "significant flow impediments" and "rely on incomplete or inaccurate information, rest in large part on assumptions that have not been validated, and are subject to far greater uncertainties than have been acknowledged." BP, BP'S PRELIMINARY RESPONSE TO THE FLOW RATE AND VOLUME ESTIMATES CONTAINED IN STAFF WORKING PAPER NO. 3 (Oct. 21, 2010).
[12] 26 U.S.C. § 9509(b)(8); 33 U.S.C. § 1321(s).
[13] EPA, SUPPLEMENTAL ENVIRONMENTAL PROJECTS POLICY 4 (Apr. 10, 1998), http://www.epa.gov/compliance
[14] EPA, BEYOND COMPLIANCE: SUPPLEMENTAL ENVIRONMENTAL PROJECTS 3-4 (Jan. 2001),
http://www.epa.gov/Compliance/resources
[15] EPA, SUPPLEMENTAL ENVIRONMENTAL PROJECTS POLICY 1 (Apr. 10, 1998).

[16] *Id.* at 2.
[17] *Id.* at 5-6.
[18] *Id.* at 6.
[19] *Id.*
[20] *Id.* at 4.
[21] *Id.* at 12-17.
[22] 33 U.S.C. §§ 1319(c)(6), 1362(5).
[23] 33 U.S.C. § 1319(c)(1).
[24] 33 U.S.C. § 1319(c)(2).
[25] 33 U.S.C. § 1319(c)(3).
[26] *Id.*
[27] 26 U.S.C. § 9509(b)(8); 33 U.S.C. § 1321(s).
[28] 18 U.S.C. § 3553(a).
[29] 18 U.S.C. § 3572(a).
[30] 18 U.S.C. § 3663a.
[31] *See, e.g.,* EXXON VALDEZ OIL SPILL TRUSTEE COUNCIL, SETTLEMENT, http://www.evostc.state
[32] 18 U.S.C. § 3572(b).
[33] 18 U.S.C. § 3571(c).
[34] U.S.S.G. § 8B1.3.
[35] *See, e.g.,* Plea Agreement, United States v. Overseas Shipholding Group, Inc., No. 06-cr-65 (E.D. Tex. Dec. 18, 2006) (the defendant pleaded guilty and agreed to a $9.2 million community service payment in addition to a $27.8 million fine).
[36] Plea Agreement, United States v. Exxon Corp., No. A90-015 CR (D. Alaska, Apr. 24, 1991); *see also* Government's Memorandum in Support of Agreement and Consent Decree, United States v. Exxon Corp., No. A91-082 CIV (D. Alaska, Oct. 8, 1991).
[37] U.S. COAST GUARD, OIL SPILL LIABILITY TRUST FUND (OSLTF) FUNDING FOR OIL SPILLS (Jan.2006), http://www.uscg.mil/npfc/docs/PDFs/OSLTF_Funding_for_Oil_Spills.pdf.
[38] 26 U.S.C. § 9509(a)-(b).
[39] Emergency Economic Stabilization Act of 2008, Pub. L. No. 110-343, § 405(a), (b), 122 Stat. 3765, 3860-61; *compare* 26 U.S.C. § 4611 (2007) *with* 26 U.S.C. § 4611 (2010).
[40] *Compare* 26 U.S.C. § 4611 (2007) *with* 26 U.S.C. § 4611 (2010).
[41] NATIONAL POLLUTION FUND CENTER, OIL POLLUTION ACT (OPA) FREQUENTLY ASKED QUESTIONS, http://www.uscg.mil/npfc/About_NPFC/opa_faqs.asp#faq1.
[42] U.S. COAST GUARD, OIL SPILL LIABILITY TRUST FUND (OSLTF) FUNDING FOR OIL SPILLS at 1.
[43] Act to Amend the Oil Pollution Act of 1990, Pub. L. No. 111-191, 124 Stat. 1278 (2010).
[44] 33 U.S.C. §§ 2712(a), 2761(f).
[45] 26 U.S.C. § 9509(c)(2).
[46] 33 U.S.C. § 2702(b)(1). The National Contingency Plan is a set of federal regulations prescribing the government's response to spills and threatened spills of oil and other hazardous materials.
[47] *Id.* § 2702(b)(2).
[48] *Id.* § 2706(d).
[49] *Id.* § 2706(f).
[50] *Id.* § 2703(a).
[51] *Id.* § 2704(a).

[52] *Id.* § 2704(c)(1).
[53] Statement of BP Exploration & Production Inc. Re Applicability of Limit of Liability Under Oil Pollution Act of 1990, In Re: Oil Spill by the Oil Rig "Deepwater Horizon" in the Gulf of Mexico, on April 20, 2010, 10-md-2179 (E.D. La. Oct. 18, 2010); *see also* The Role of BP in the Deepwater Horizon Explosion and Oil Spill: Hearing before the Subcomm. On Oversight and Investigations of the H. Comm. on Energy and Commerce, 111th Cong. (2010) (written testimony of Tony Hayward, CEO of BP) (BP has committed to paying "all necessary clean up costs and all legitimate claims for other losses and damages caused by the spill.").
[54] BP, CLAIMS AND GOVERNMENT PAYMENTS GULF OF MEXICO SPILL PUBLIC REPORT (Dec. 22, 2010), http://www.bp.com/liveassets/bp_internet/globalbp/globalbp_uk_english/incident_response downloads_pdfs/Public_Report_12.22.10.pdf.
[55] *Id.* § 2718.
[56] BP, CLAIMS AND GOVERNMENT PAYMENTS GULF OF MEXICO SPILL PUBLIC REPORT (Feb. 10, 2011), http://www.bp.com/liveassets/bp_internet/globalbp/globalbp_uk_english/incident_response downloads_pdfs/Public_Report_2-10-11.pdf.
[57] A Congressional Research Service report discussing criminal liability related to the spill under wildlife laws concluded that it would be difficult for the government to bring successful criminal prosecutions under the Endangered Species Act and the Marine Mammal Protection Act because of the *mens rea* requirements of those acts. By contrast, according to the report, it would be easier for the government to successfully bring criminal prosecutions under the misdemeanor provisions of the Migratory Bird Treaty Act, which imposes strict liability on defendants. CONGRESSIONAL RESEARCH SERVICE, THE 2010 OIL SPILL: CRIMINAL LIABILITY UNDER WILDLIFE LAWS (June 28, 2010), available at: http://assets. The Congressional Research Report does not discuss civil liability.
[58] 16 U.S.C. § 1538(a). Take means to "harass, harm, pursue, hunt, shoot, wound, kill, trap, capture, or collect, or to attempt to engage in any such conduct." *Id.* § 1532(19).
[59] 16 U.S.C. § 1540(a); 15 C.F.R. § 6.4(e)(13).
[60] 16 U.S.C. § 1540(b).
[61] *Id.* § 1540(a)(1); 15 C.F.R. § 6.4(e)(13).
[62] 16 U.S.C. § 1540(d).
[63] *Id.* §§ 1535, 1540(d).
[64] *Id.* § 1372(a).
[65] *Id.* § 1375(b).
[66] *Id.* § 1375(a)(1); 15 C.F.R. § 6.4(e)(10).
[67] Take means to "harass, harm, pursue, hunt, shoot, wound, kill, trap, capture, or collect, or to attempt to engage in any such conduct." 16 U.S.C. § 1532(19). And "harassment" includes "any act of pursuit, torment, or annoyance which (i) has the potential to injure a marine mammal or marine mammal stock in the wild; or (ii) has the potential to disturb a marine mammal or marine mammal stock in the wild by causing disruption of behavioral patterns including, but not limited to, migration, breathing, nursing, breeding, feeding, or sheltering." *Id.* § 1362(18)(a). *See also* CONGRESSIONAL RESEARCH SERVICE, THE 2010 OIL SPILL: CRIMINAL LIABILITY UNDER WILDLIFE Laws at 4.
[68] 16 U.S.C. § 1375a.
[69] *Id.* § 703(a).

[70] *Id.* § 707(a); *see also, e.g.*, United States v. Boynton, 63 F.3d 337. 343 (4th Cir. 1995) (holding that strict liability is the standard).

[71] 16 U.S.C. § 707(b)-(c).

[72] *See* CONGRESSIONAL RESEARCH SERVICE, THE 2010 OIL SPILL: CRIMINAL LIABILITY UNDER WILDLIFE LAWS at 7- 8.

[73] *See* Exxon Shipping Co. v. Baker, 554 U.S. 471, 479 (2008) (recounting Exxon's settlement with the United States for criminal violations of the Clean Water Act, the Refuse Act of 1899, and the Migratory Bird Treaty Act); *see also* CONGRESSIONAL RESEARCH SERVICE, THE 2010 OIL SPILL: CRIMINAL LIABILITY UNDER WILDLIFE LAWS at 7-8.

[74] 42 U.S.C. § 10601.

[75] 16 U.S.C. § 4406.

In: Gulf Oil Spill of 2010... ISBN: 978-1-61324-729-7
Editors: C. R. Walsh, J. P. Duncan © 2012 Nova Science Publishers, Inc.

Chapter 8

THE 2010 OIL SPILL: CRIMINAL LIABILITY UNDER WILDLIFE LAWS[*]

Kristina Alexander

SUMMARY

The United States has laws that make it illegal to harm protected wildlife. Those laws could be used to prosecute those who caused the 2010 oil spill. Perhaps the most famous of these laws is the Endangered Species Act (ESA), which provides for both criminal and civil penalties for acts that harm species listed under the act. The Marine Mammal Protection Act (MMPA) also provides for civil and criminal punishment when an action takes a marine mammal. The Migratory Bird Treaty Act (MBTA) makes it a crime to kill migratory birds.

While there are endangered species and marine mammals in the area affected by the Gulf of Mexico oil spill, it is more likely that any criminal prosecution would use the MBTA rather than the ESA or the MMPA. This is because the MBTA is a strict liability statute in relevant part, unlike the other laws. Accordingly, the prosecution does not have to show that the defendant(s) intended to harm wildlife. The prosecution does not have to prove that the defendants knew their action(s) would lead to an oil spill to find liability. The MBTA was used to prosecute Exxon

---

[*] This is an edited, reformatted and augmented version of a Congressional Research Service publication, CRS Report for Congress R41308, from www.crs.gov, dated August 31, 2010.

following the *Exxon Valdez* spill and has been used for decades to find corporations and even their employees criminally liable for the deaths of protected birds.

INTRODUCTION

In April 2010 an explosion occurred on an oil rig in the Gulf of Mexico, reportedly killing 11 people, and, causing the worst oil spill in U.S. history.[1] Millions of barrels of oil are believed to have leaked into the Gulf of Mexico. As the oil spreads, the implications for harm to wildlife grow.

The United States has many laws that protect wildlife from harm. This report will discuss three: the Endangered Species Act, the Marine Mammal Protection Act, and the Migratory Bird Treaty Act.[2] The Endangered Species Act (ESA), for example, prohibits actions that harass, harm, wound, or kill a listed species (meaning either a species listed as threatened or endangered).[3] The Marine Mammal Protection Act (MMPA) prohibits actions that harass or kill marine mammals, further defining harassment as "any act of pursuit, torment, or annoyance which (i) has the potential to injure a marine mammal ... or (ii) has the potential to disturb a marine mammal ... by causing disruption of behavioral patterns."[4] The Migratory Bird Treaty Act (MBTA) makes it unlawful "by any means or in any manner" to take, kill, or attempt to take or kill "any migratory bird, any part, nest, or egg of any such bird."[5]

CRIMINAL LAW BASICS

Jurisdiction of U.S. Laws on the Outer Continental Shelf

Before analyzing what behavior may be a crime under those laws, some initial jurisdictional issues need to be discussed. The wildlife statutes of this report have different rules regarding where they apply. For instance, the ESA applies to "persons under the jurisdiction of the United States" who take a listed species in the United States, on the territorial seas of the United States, or on the high seas.[6] The MMPA applies to the United States as well as "waters under the jurisdiction of the United States," which is defined as including territorial sea, and waters 200 miles seaward from its coast.[7] The MBTA has no statement regarding its jurisdictional reach.

With or without a statement of jurisdiction, these acts apply to the oil spill from a well located approximately 50 miles off the coast of Louisiana. The Outer Continental Shelf Lands Act (OCSLA) attaches U.S. jurisdiction over the site:

> The Constitution and laws and civil and political jurisdiction of the United States are extended to the subsoil and seabed of the outer Continental Shelf and to all artificial islands, and all installations and other devices permanently or temporarily attached to the seabed, which may be erected thereon for the purpose of exploring for, developing, or producing resources therefrom, or any such installation or other device (other than a ship or vessel) for the purpose of transporting such resources, to the same extent as if the outer Continental Shelf were an area of exclusive Federal jurisdiction located within a State.[8]

Jurisdiction over the People Who Committed the Acts

The next issue in enforcement is to find to whom these laws apply. Each describes a violation in terms of its being committed by a *person*. Under the ESA, a *person* is defined as including an individual, corporation, and private entity, and an officer, employee, agent of the federal, a state, or foreign government.[9] The MMPA defines *person* to include people, businesses, and parts of governments, such as employees, departments, and agents.[10] While the MBTA does not define *person,* it states that "any person, association, partnership, or corporation" may be guilty of a crime under the act.[11] Accordingly, it appears that all of these laws apply not just to individual actions but to those of corporations or other business entities, or people acting in an official capacity for those organizations.

PROSECUTION UNDER THE WILDLIFE STATUTES

A violation of a law may be either civil or criminal. The ESA and the MMPA provide for both criminal and civil violations, while the MBTA addresses only criminal prosecutions. Civil and criminal violations differ in how they were committed and how they are punished. This report will focus on criminal violations.

Endangered Species Act

Typically, there are differences between the type of behavior that leads to a civil violation versus the type of behavior that leads to a criminal violation. However, the ESA makes no distinction, requiring only that the violation be done *knowingly*. Under the ESA, someone has committed a violation of the ESA if they "knowingly violate ... any provision of this chapter, or any provision of any permit or certificate issued hereunder."[12] That language appears both in Section 1340(a) for civil violations and Section 1340(b) for criminal violations. It is up to the prosecutor (or the agency recommending prosecution) whether the action is charged civilly or criminally. In either case, killing or harming a listed species violates the act (if it is not otherwise permitted under the act).[13] A criminal violation of the ESA is a misdemeanor.

The term *knowing* applies to the act being committed, not to the violator's understanding of the law; that is, a hunter does not have to know that the animal she shot was endangered, only that she knew she was shooting something.[14] It refers to the defendant's mindset, or *mens rea*. In contrast, other statutes may require *specific intent* to commit an illegal act, meaning the defendant intended to commit a crime. Instead, the knowing standard is described as requiring *general intent*, meaning the defendant intended to commit an act and that act turned out to be a crime. The ESA has been held to be a general intent statute.[15]

The ESA used to be a specific intent statute. Before being amended, the statute required a prosecutor to show the defendant acted *willfully* to establish a criminal violation.[16] One environmental law expert has noted how the knowing standard of mens rea over other, stricter, requirements can make prosecution easier:

> Why is this more relaxed mens rea element important for the government's deterrence objectives? ... Relaxing mens rea ... can dramatically improve the prosecutor's chance of success. This is true for crime in general. Indeed, it is especially so for environmental crime, at least where those violating environmental regulations are large corporations where individuals making decisions may seek to remain willfully ignorant.[17]

ESA and the Oil Spill

The knowing requirement seemingly nullifies any prosecution for the oil spill for harm to listed species. A crime, or in fact, a civil violation, could be prosecuted under the ESA only if someone (e.g., BP, Deepwater Horizon,

Halliburton, or any of their employees or agents) *knew* that their actions would lead to an oil spill. Under a general intent statute, a defendant has to intend to commit an act (meaning the spill or the blowout) and that act turned out to be a crime. That is a difficult standard to prove in this context. It is not enough to show that someone was reckless or irresponsible, or ignored important information. To convict, it must be shown that the defendant or defendants knew their actions would cause a spill.

Marine Mammal Protection Act

Because the MMPA also has a knowing standard for its criminal violations, it appears that criminal prosecution under this act for harm to marine mammals such as dolphins, whales, and manatees, is unlikely.

Under the MMPA, a "person who knowingly violates any provision of this subchapter or of any permit or regulation issued thereunder [not including takings by commercial fishing operations] shall, upon conviction, be fined not more than $20,000 for each such violation, or imprisoned for not more than one year, or both."[18] The case law for the MMPA is not as extensive as that of the ESA. However, the Ninth Circuit has discussed the mens rea under the MMPA, clarifying that the standard is *knowing*, not merely *negligence*.[19] Like the ESA, violations of the MMPA are misdemeanors.

While the ESA requires establishing knowing conduct to prove a civil violation, the MMPA has no such requirement. Civil violations occur if a person "violates any provision of this subchapter or of any permit or regulation issued thereunder."[20] Accordingly, intent is not necessary. A civil fine under the MMPA can reach $11,000.[21]

MMPA and the Oil Spill

Thus, while the ESA does not appear to provide a vehicle for either civil or criminal prosecution for animals killed or injured by the oil, due to its knowing standard, and criminal prosecutions under the MMPA are similarly restricted, civil charges remain available under that act. Civil violations may be fined up to $11,000 for each violation.[22] Because the MMPA penalizes activities that not only kill or injure a marine mammal, but those that harass or interrupt breeding, breathing, feeding, or sheltering, the fine could be substantial.

Migratory Bird Treaty Act

The mens rea for the MBTA is different from that in the ESA and the MMPA. It allows prosecution of misdemeanors based on strict liability.[23] This means instead of having to show that defendants knew they were committing a particular act, the prosecution only has to show that an act happened. Under strict liability, actors are liable for a violation regardless of what they knew or what they meant to do. It is the easiest standard under which to prosecute.

The portion of the MBTA that addresses misdemeanor violations[24] invokes strict liability: "any person, association, partnership, or corporation who shall violate any provisions of said conventions or of this subchapter, or [violate a regulation] shall be deemed guilty of a misdemeanor and upon conviction thereof shall be fined not more than $15,000 or be imprisoned not more than six months, or both."[25] The statute is expansive, prohibiting taking a migratory bird "by any means or in any manner."[26] *Taking* is defined in the regulations as "pursue, hunt, shoot, wound, kill, trap, capture, or collect."[27]

The statutory history of the MBTA makes reference to its strict liability provisions. In discussion regarding the 1986 amendment of the act, in which the felony provision regarding taking a bird to sell or barter was modified, the Senate addressed the misdemeanor provisions that are at issue here:

> Nothing in this amendment is intended to alter the "strict liability" standard for misdemeanor prosecutions under 16 U.S.C. 707(a), a standard which has been upheld in many Federal court decisions.[28]

In 1998, the Senate again discussed preserving the strict liability provision for misdemeanors even though the provision that addressed the mens rea for bird baiting[29] violations was being modified. It focused on the long history of strict liability under the act:

> The elimination of strict liability, however, applies only to hunting with bait or over baited areas, and is not intended in any way to reflect upon the general application of strict liability under the MBTA. Since the MBTA was enacted in 1918, offenses under the statute have been strict liability crimes. The only deviation from this standard was in 1986, when Congress required scienter for felonies under the Act.[30]

Although the original purpose of the MBTA was to protect birds from hunting, it has long been used to prosecute any sort of taking of birds, including when they die from contamination. The Department of Justice, in a memorandum supporting its prosecution of an oil company and its employee for killing migratory birds in open-top oil tanks,[31] refers to three unpublished cases in which oil companies were convicted under the MBTA for the deaths of birds in oil sump pits.[32] That same DOJ Memorandum refers to three cases in which companies and/or individuals were convicted of killing migratory birds as a result of contaminated mining tailings or collection ponds.[33] These cases illustrate that convictions under the MBTA are possible when migratory birds are killed incidentally by being poisoned or otherwise harmed by contaminants, rather than being specifically targeted.

In one case, the Second Circuit discussed the strict liability standard of the MBTA. The court referenced the jury instructions of the trial court, in which jurors were advised that:

> under the law good will and good intention and measures taken to prevent the killing of the birds are not a defense.... The Government in this case does not have to prove that the defendant intended to kill the birds. You may convict the corporation even if you find that the killing of the birds was accidental or unintentional provided that you find that the FMC Corporation did kill the birds as charged in the indictment.[34]

In that case, a pesticide manufacturer was charged with killing 92 migratory birds that died in a wastewater pond at its plant. The Second Circuit found that no intent to kill the birds was necessary to support a conviction under the MBTA.[35] Its decision rested in part on the fact that the defendant was manufacturing something known to be dangerous. It distinguished the criminal act of the defendant from other actions, such as accidently hitting a bird while driving, stating that any punishment for those technical violations is best left to "the sound discretion of prosecutors and the courts."[36] Under a similar analysis, it appears likely that an oil spill, which is known to be toxic, would be an act for which a defendant may be found strictly liable under the MBTA.

The Tenth Circuit revisited the constitutionality of the strict liability standard.[37] It upheld the statute but found that, under certain circumstances, the application of the act could be unconstitutional. The court was reviewing the conviction of two operators of oil drilling equipment known as heater-treaters. Dead birds had been found in heater-treaters in the area, and the U.S. Fish and Wildlife Service (FWS) had conducted an educational campaign to inform operators of the danger and liability. Following the educational campaign, two different operators were prosecuted for migratory birds killed in their equipment. The Tenth Circuit reversed one of the convictions, finding that the death had occurred before the operator had knowledge that the behavior could be criminal. The court said that when actions are commonly not criminal or dangerous, there must be fair notice before strict liability can be imposed.[38] This may be distinguished from the facts of the Gulf oil spill in that oil is known to be toxic.

Another court also questioned the strict liability standard in the oil patch. Despite the acceptance by the Fifth Circuit of the strict liability standard for MBTA crimes,[39] a lower federal court in Louisiana refused to impose strict liability on an oil company whose equipment caused the death of 14 brown pelicans. In *United States v. Chevron USA, Inc.*, the court for the Western District of Louisiana rejected a plea agreement in which Chevron pleaded guilty to having killed the birds in violation of the MBTA.[40] The district court referenced Fifth Circuit precedent,[41] noting the Fifth Circuit had been in the minority in refusing to accept strict liability to that class of criminal cases (baiting) before the law was amended.[42] The district court then indicated it would not accept strict liability for this type of crime. The district court stated that it "declines to extend the Fifth Circuit's application of strict liability under the MBTA" to the conduct of the defendant. The court found the defendant did not have "fair warning" that its drilling rigs would kill brown pelicans. In addition to rejecting Fifth Circuit precedent, which it found was unrelated, the court rejected the persuasive value of the convictions in the *Apollo* case by the lower Kansas federal court (before the Tenth Circuit had decided the appeal),[43] saying it "simply declines to adopt the reasoning and rationale of that decision."[44]

MBTA and the Oil Spill

The most similar case to the 2010 oil spill is that of the *Exxon Valdez*, in which Exxon Corp. and a subsidiary were convicted of violations of the MBTA.[45] Exxon agreed to a $150 million criminal fine, but as part of an agreement with the United States, paid $25 million instead.[46] Exxon also paid $100 million in criminal restitution for injuries to fish, wildlife, and habitat.[47]

Based on other similar prosecutions under the MBTA, a court could possibly find liability for the corporations involved in the 2010 spill, their subsidiaries, and their officers and employees, which, under the MBTA, includes fines and up to six months imprisonment.[48]

There is no way at this point to estimate what, if any, criminal penalty might be assessed against a defendant as a result of the oil spill. According to FWS, 5,362 bird carcasses have been collected as of August 30, 2010, although that number cannot account for the number of birds that simply sunk after being coated with oil.[49] For example, following the *Exxon Valdez* spill, 35,000 dead birds were found, but the Exxon Valdez Oil Spill Trustee Council estimated that over 250,000 birds were killed.[50] Additionally, it does not always follow that an MBTA fine is calculated per dead bird. At least one court has held that the crime is based on the number of acts that led to birds being killed and not on the number of birds killed.[51] The act also makes it a crime to *take* a bird, which includes wounding it. It is not clear whether being coated in oil would be considered a wound.

MBTA criminal penalties reported in case law are varied. In the *United States v. FMC Corp.*, summarized above, 92 bird deaths resulted in a $500 penalty.[52] At the time of that case, the maximum fine was $500. In a 2009 case in which two migratory birds were found dead in machines used to treat crude oil, a total of $2,000 in fines was assessed: $1,500 against a corporation for one violation, and $250 against an owner/operator for one violation.[53] A steel company was assessed a $65,900 fine for the deaths of 28 migratory birds in its runoff water system, although ultimately found not guilty.[54]

Alternative Fines Act

Prosecution under the MBTA may also consider a section of the criminal code, 18 U.S.C. § 3571, which originated in the Comprehensive Crime Control Act of 1984 and was amended by the Alternative Fines Act. The Alternative Fines Act addressed certain problems with criminal penalties. Congress created the law to "make criminal fines more severe."[55] The statute relates penalties to the incarceration time and sets uniform fines for those felonies and misdemeanors. It does this in two steps: one to classify the crime (18 U.S.C. § 3559); and the other to set the fines based on those classifications (18 U.S.C. § 3571). Thus, under 18 U.S.C. § 3559, a statute that imposes no more than six months in jail (such as the MMPA and the MBTA) is classified as a Class B misdemeanor. A statute that imposes no more than one year imprisonment (such as the ESA) is classified as a Class A misdemeanor. The fine amounts for Class A and Class B misdemeanors are set under 18 U.S.C. § 3571. A Class A misdemeanor imposes a maximum fine of $100,000 on an individual defendant, and a Class B misdemeanor imposes a maximum fine of $5,000 for an individual defendant. The fines in both cases are doubled for defendants that are organizations.

However, if a statute's penalty provision post-dates that 1984 act, the fines under Section 3571 do not apply. Both the ESA and the MBTA were amended to change the criminal penalties after 1984.[56]

Congress changed the ESA's criminal fines in 1988.[57] The amendment set the maximum criminal fine at $50,000. Even if prosecution were brought under the ESA, which, as discussed above, appears unlikely, the $100,000 criminal penalty provided by Section 3571 for a Class A misdemeanor would not apply.[58]

Similarly, the 1998 amendment of the MBTA set a fine of $15,000.[59] However, this fine is larger than the $5,000 provided by Section 3571 for a Class B misdemeanor.

The Alternative Fines Act also includes a doubling provision that allows a fine of twice any financial loss: "if the offense results in pecuniary loss to a person other than the defendant, the defendant may be fined not more than the greater of twice the gross gain or twice the gross loss."[60] However, it is likely that a court would find that the MBTA amount prevails and the doubling provision is not available. For one thing, the current MBTA criminal fine provision postdates the Alternative Fines Act. Additionally, when amending the MBTA fines, Congress specifically applied the Alternative Fines Act to another type of violation (the bird baiting provision in Section 707(c)), suggesting a purposeful omission regarding the misdemeanor provision. If a court held that it did apply, defendants prosecuted for the oil spill could be responsible not just for the $15,000 fine under the MBTA, but for twice the pecuniary losses incurred by anyone as a result of the spill, such as cleanup expenses and lost revenue.[61]

CONCLUSION

While the three major wildlife statutes provide for criminal penalties, the mens rea requirement of two of them (the ESA and the MMPA) appears to prohibit their application to the 2010 oil spill. Both require proof that a defendant knew it was committing the act of causing an oil spill, which does not appear at this point to match the known circumstances surrounding the oil spill. However, the MBTA imposes strict liability on defendants and seems to be a useful tool for prosecution, especially in light of what appear to be massive numbers of bird deaths as a result of the oil spill. Strict liability means that the prosecution has to show only that the defendant was responsible for the act, not that it knew or intended to spill oil.

Earlier prosecutions under the MBTA have charged corporations, their subsidiaries, and even employees with crimes under the MBTA, making it likely that the same type of defendants could be named should the United States prosecute under the act. Criminal fines of up to $15,000 per act are available under the law; however, it is not known whether that fine would be assessed per dead or wounded migratory bird or per criminal action.

ACKNOWLEDGMENTS

The author would like to thank Charles Doyle of the American Law Division for sharing his expansive criminal law expertise.

End Notes

[1] An estimated total of 4.1 million bbl were released into the Gulf (4.9 million bbl leaked, but 800,000 bbl captured before it leaked into the Gulf). See Official Site of the Deepwater Horizon Unified Command, at www.deepwaterhorizonresponse.com/go/doc/2931/840475.

[2] Potential criminal liability under other federal and state criminal laws is beyond the scope of this report. For information on how oil may harm wildlife and their habitat, see CRS Report R41311, *The Deepwater Horizon Oil Spill: Coastal Wetland and Wildlife Impacts and Response*, by M. Lynne Corn and Claudia Copeland.

[3] 16 U.S.C. § 1538—prohibiting taking listed species; 16 U.S.C. § 1532—defining take as "harass, harm, pursue, hunt, shoot, wound, kill, trap, capture, or collect, or to attempt to engage in any such conduct."

[4] 16 U.S.C. §§ 1362(13), 1362(18A).

[5] 16 U.S.C. § 703(a).

[6] 16 U.S.C. § 1538(a).

[7] 16 U.S.C. § 1362(15).

[8] 43 U.S.C. § 1333(a)(1).

[9] 16 U.S.C. § 1532(13).

[10] 16 U.S.C. § 1362(10).

[11] 16 U.S.C. § 707(a).

[12] 16 U.S.C. § 1340.

[13] The Minerals Management Service (MMS) (now known as the Bureau of Ocean Energy Management, Regulation, and Enforcement [BOEMRE]) indicated that the following endangered species are likely to be in the area of the spill: Northern Right whale, Blue whale, Fin whale, Sei whale, Humpback whale, Sperm whale, West Indian manatee, Leatherback turtle, Green turtle, Hawksbill turtle, Kemp's Ridley turtle, Loggerhead turtle, Gulf sturgeon, Whooping crane, Piping plover, Alabama beach mouse, Choctawhatchee beach mouse, St. Andrew beach mouse, and Perdido Key beach mouse. MMS Outer Continental Shelf Oil & Gas Leasing Program: 2007-2012, Final Environmental Impact

Statement MMS 2007-003. Available at http://www.boemre.gov/5-year/2007-2012FEIS.htm.

[14] See United States v. McKittrick, 142 F.3d 1170, 1176-77 (9th Cir. 1998) (gray wolf) (holding intent requirement of ESA was satisfied when defendant knew he shot an animal, and that animal turned out to be a protected gray wolf, not that the defendant knew that the animal was protected); United States v. St. Onge, 676 F. Supp. 1044, 1045 (D. Mont. 1988) (grizzly bear) (indicating that issue is whether defendant shot at an animal knowingly, not whether defendant recognized the animal he was shooting); United States v. Billie, 667 F. Supp. 1485, 1492 (S.D. Fla. 1987) (Florida panther) (rejecting defendant's argument that government needs to prove defendant knew animal he was shooting was protected as "without support in law or reason").

[15] See U.S. v. Kapp, 2003 WL 23162408, *8 (N.D. Ill. 2003) ("the ESA does not require proof that the defendant had a criminal objective or was even aware of the unlawfulness of his conduct").

[16] P.L. 95-632, § 6(4) substituted *knowingly* for *willingly* in what is codified at Section 1540(b)(1), addressing criminal violations. The *knowing* requirement in the civil provision (Section 1540(a)) appeared in the original public law and remains.

[17] Richard J. Lazarus, *Mens Rea In Environmental Criminal Law: Reading Supreme Court Tea Leaves*, 7 Fordham Envtl. L.J. 861 (Symposium, 1996).

[18] 16 U.S.C. § 1375(b).

[19] United States v. Hayashi, 22 F.3d 859, 862 (9th Cir. 1994) ("Under the MMPA, no criminal penalty can attach for negligent conduct").

[20] 16 U.S.C. § 1375(a)(1).

[21] According to the penalty chart posted by the General Counsel of the National Oceanic and Atmospheric Administration, the recommended minimum fine for killing a marine mammal is $3,250 for a first violation, $4,500 for a second violation, and $8,500 for a third violation. See NOAA Office of General Counsel Penalty Schedules at http://www.gc.noaa.gov/enforce-office3.html.

[22] The statute provides a civil penalty of $10,000. 16 U.S.C. § 1375(a). However, under the Debt Collection Improvement Act, that amount has been adjusted for inflation and is now $11,000. 73 Fed. Reg. 75321 (Dec. 11, 2008); 15 C.F.R. § 6.4(e)(10).

[23] A knowing standard applies to felony charges relating to taking migratory birds with the intent to sell or selling migratory birds. 16 U.S.C. § 707(b). Additionally, charges relating to baiting birds require proof of knowledge. 16 U.S.C. § 704(b).

[24] One section of the MBTA has a felony provision but pertains to selling birds. 16 U.S.C. § 707(b). Another portion of the MBTA requires a knowing violation but pertains to baiting birds. 16 U.S.C. § 704(b). Those two sections are not relevant to this discussion.

[25] 16 U.S.C. § 707(a).

[26] 16 U.S.C. § 703(a).

[27] 50 C.F.R. § 10.12.

[28] S. Rep. 99-445 (1986), 1986 U.S.C.C.A.N. 6113, 6128.

[29] Baiting birds is when hunters put food in an area to draw birds to them. Violations related to baiting are found in 16 U.S.C. § 704(b).

[30] S. Rep. 105-366, at 3 (1998). One court rejected this history as being commentary of a later Congress rather than evidence of what a current Congress has intended to legislate. U.S. v. Chevron USA, Inc., No. 09-CR-0132, 2009 WL 3645170 (W.D. La. 2009) (rejecting a plea agreement wherein Chevron had agreed to plead guilty under the MBTA to killing 14 brown pelicans trapped in its rig caisson).

[31] United States v. Citgo Petroleum Corp., No. 06-563-S, 2007 WL 2049382, Trial Memorandum of the United States Regarding Violations of the Migratory Bird Treaty Act - Counts Six Through Ten (S.D. Texas June 26, 2007) (hereinafter *DOJ Memorandum*).

[32] DOJ Memorandum at 3 (United States v. Stuarco Oil Co., 73-CR-129 (D. Colo. 1973) (company charged and pled nolo contendere to 17 counts under the MBTA for deaths of birds resulting from company's failure to build oil sump pits in a manner that could keep birds away); United States v. Union Tex. Petroleum, 73-CR-127 (D. Colo. 1973) (prosecution of oil company for maintenance of oil sump pit); United States v. Equity Corp., Cr. 75-51 (D. Utah 1975) (company charged with and pleaded guilty to 14 counts under the MBTA for deaths of 14 ducks caused by the company's oil sump pits)).

[33] DOJ Memorandum at 3 (United States v. Kennecott Communications Corp., No. N-90-16M (D. Nev. Mar. 8, 1990); United States v. Echo Bay Minerals Co., No. CR N-90-52-HDM (D. Nev. 1990); United States v. Nerco-Delamar Co. (a.k.a. Delamar Silver Mine), No. CR 91-032-S-HLR (D. Idaho Apr. 21, 1992)).

[34] United States v. FMC Corp., 572 F.2d 902, 904 (2d Cir. 1978) (the $500 fine under the MBTA was upheld).

[35] For other circuit court decisions upholding strict liability for MBTA violations (although not necessarily related to contamination), see United States v. Morgan, 311 F.3d 611 (5th Cir. 2002); United States v. Corrow, 119 F.3d 796 (10th Cir. 1997); United States v. Boynton, 63 F.3d 337 (4th Cir. 1995); United States v. Engler, 806 F.2d 425 (3d Cir. 1986); United States v. Wood, 437 F.2d 91 (9th Cir. 1971).

[36] United States v. FMC Corp., 572 F.2d at 905.

[37] United States v. Apollo Energies, Inc., 611 F.3d 679 (10th Cir. 2010).

[38] United States v. Apollo Energies, Inc., 611 F.3d 679, *18-19 (10th Cir. 2010).

[39] United States v. Stephans, 142 Fed. Appx. 821 (5th Cir. 2005); United States v. Morgan, 311 F.3d 611 (5th Cir. 2002).

[40] No. 09-CR-0132, 2009 WL 3645170 (W.D. La. Oct. 30, 2009) (*Chevron USA*).

[41] United States v. Sylvester, 848 F.2d 520 (5th Cir. 1988) (holding that prosecution could not show that hunters knew the field was baited and that strict liability did not apply); United States v. Delahoussaye, 573 F.2d 910 (5th Cir. 1978) (holding that penalizing hunters when they could not have known field was baited was contrary to law).

[42] Chevron USA, at *3.

[43] United States v. Apollo Energies, Inc., Nos. 08-10111-01-JTM, 08-10112-01-JTM, 2009 WL 211580 (D. Kan. January 28, 2009), *aff'd in part, rev'd in part*, United States v. Apollo Energies, Inc., 611 F.3d 679 (10th Cir. 2010).

[44] Chevron USA, at *5.

[45] United States v. Exxon Corp., No. A90-015 CR (D. Alaska sentencing April 24, 1991). While the civil penalty challenge was resolved by the U.S. Supreme Court (*Exxon Shipping Corp. v. Baker*, 128 S. Ct. 2605 (2008)), no decision regarding the criminal penalties was published.

[46] Exxon Valdez Oil Spill Trustee Council Settlement webpage at http://www.evostc.state. See also Exxon Shipping Corp. v. Baker, 128 S. Ct. at 2613 (referring to the criminal prosecution).

[47] Id. The misdemeanors under the MBTA were just one of five different types of criminal charges brought, which was the only wildlife act used.

[48] The defendants for the *Exxon Valdez* spill included Exxon and a subsidiary. The ongoing case against Citgo, referenced above, also includes a subsidiary and the environmental manager of the site as defendants. United States v. Citgo Petroleum Corp, C-05-563 (S.D. Texas).

See also United States v. Cota, No. CR 08-00160 (N.D. Cal. June 22, 2009) (accepting the plea of a ship's captain under the Clean Water Act and the MBTA for harm resulting from oil spill following allision with the San Francisco Bay Bridge; management company of the ship was also a defendant).

[49] FWS, Consolidated Fish and Wildlife Collection Report (August 30, 2010), at http://www.fws.gov/home/dhoilspill/collectionreports.html.

[50] http://www.evostc.state

[51] United States v. Corbin Farm Service, 444 F. Supp. 510 (E.D. Cal. 1978).

[52] United States v. FMC Corp., 572 F.2d 902 (2d Cir. 1978).

[53] United States v. Apollo Energies, Inc., 611 F.3d 679 (10th Cir. 2010).

[54] United States v. WCI Steel, Inc., No. 5:04 MJ 5053, 2006 WL 2334719 (N.D. Ohio August 10, 2006).

[55] H.R. Rep. 98-906, at 1 (1984), 1984 U.S.C.C.A.N. 5333, 5433.

[56] The original criminal penalty of the MMPA appears to have been superseded by Section 3571. However, the knowing standard is likely to bar criminal prosecution under this act, and so this act is not discussed further.

[57] P.L. 100-478, § 1007, 102 Stat. 2309.

[58] See United States v. Eisenberg, 496 F. Supp. 2d 578. 581-3 (E.D. Pa. 2007) (holding that the ESA amendments that post-date the 1984 act evidence congressional intent to impose a criminal fine lower than that of the 1984 act).

[59] P.L. 105-312, § 103, 112 Stat. 2956.

[60] 18 U.S.C. § 3571(d).

[61] The Alternative Fines Act was used to assess the criminal fine against Exxon. However, that prosecution pre-dated the 1998 change to the MBTA, meaning the MBTA amendment superseding the Alternative Fines Act was not in place. Also, the grouping of the fines in the Exxon prosecution makes it impossible to determine whether any doubling was related to the MBTA rather than to the other criminal charges brought against the company.

INDEX

A

access, 20, 71, 88
Ad Hoc Subcommittee on Disaster Recovery, 32
adjoining shorelines, viii, 5, 22, 35, 36, 37, 38, 77
adjustment, 11, 142
adults, 101
adverse conditions, 131
adverse effects, 127
age, 33
agencies, 2, 4, 11, 28, 29, 78, 80, 91, 97, 102, 103, 114, 127, 140, 146, 150
Alaska, 3, 10, 22, 23, 24, 25, 26, 27, 29, 31, 46, 111, 114, 130, 141, 149, 150, 156, 171
American Samoa, 51
appropriations, 21, 127
assessment, ix, 18, 21, 41, 67, 78, 83, 85, 86, 89, 90, 91, 96, 97, 98, 99, 101, 102, 103, 104, 107, 109, 111, 112, 114, 136, 139
assets, 157
audit, 18
authorities, 4, 144, 153
authority, 7, 18, 21, 25, 44, 46, 59, 63, 77, 93, 97, 98, 113, 114, 128, 146, 154
average costs, 142

B

background information, 62
ban, 80
bankruptcy, viii, 48, 49, 62
barter, 153, 164
base, 8, 135
benefits, 70, 88
biodegradation, 110, 115
biotic, 23
birds, ix, 78, 81, 84, 91, 101, 103, 107, 109, 111, 112, 114, 153, 159, 160, 164, 165, 166, 167, 170, 171
blame, 82
bonds, 63, 67
bounds, 139
BP Deepwater Horizon Oil Spill Commission, viii, 95
breathing, 157, 163
breeding, 157, 163
Bureau of Labor Statistics, 29
businesses, 16, 18, 68, 75, 87, 161

C

canals, 104
carbon, 115
carbon dioxide, 115
case law, 163, 167

Index

catastrophic spill, vii, 1, 2, 12, 13, 20, 119, 120, 123, 128, 137, 142
causation, 19
CBS, 32
CERCLA, 7, 77, 97, 113, 114
certificate, 55, 63, 125, 162
certification, 71
challenges, 15, 99, 103, 136
chemical, 109, 110, 139
citizens, 151, 155
civil penalties, ix, 15, 29, 144, 145, 155, 159
Clean Water Act (CWA), vii, 1, 3, 77
cleaning, 118, 119, 120, 122, 124, 139, 148
cleanup, vii, 1, 3, 6, 7, 21, 22, 23, 85, 88, 101, 113, 118, 119, 120, 123, 126, 127, 131, 132, 133, 137, 142, 168
cleanup activities, vii, 1, 6, 101
Coast Guard, ix, 7, 9, 25, 26, 28, 29, 30, 33, 51, 57, 58, 59, 82, 89, 93, 94, 98, 113, 117, 118, 120, 121, 122, 123, 124, 125, 127, 132, 134, 135, 138, 139, 141, 142, 145, 149
commerce, 50
commercial, 93, 104, 123, 126, 133, 163
common law, 76
communication, 30, 31, 93
communities, 81, 84, 106, 146
community, 46, 148, 154, 156
community service, 148, 154, 156
compensation, vii, viii, 1, 2, 3, 4, 8, 9, 10, 11, 12, 15, 18, 20, 21, 29, 30, 38, 41, 42, 43, 45, 57, 58, 62, 63, 65, 71, 76, 77, 78, 80, 86, 93, 122, 124, 127, 140
compensation framework, vii, 1, 3, 4, 9, 10, 21
competition, 68
complement, 21
compliance, 48, 52, 70, 146, 155
conference, 102
Conference Report, 81, 88
consensus, 80
consent, 78, 112, 155
conservation, 152, 154
constituents, 110
Constitution, 161

construction, 7, 39, 54, 63, 66, 122, 124, 134, 151
consulting, 84, 139
consumer price index, 8, 26
Consumer Price Index, 29, 56, 135, 141
consumers, 148
consumption, 44, 56
contaminant, 102
contamination, 76, 77, 132, 133, 140, 164, 171
Continental, 46, 47, 55, 160, 161, 169
controversies, 57
conviction, 147, 163, 164, 165
cooperation, 43, 54, 81, 97, 148
coordination, 101
coral reefs, 81, 84
cost, ix, 7, 12, 14, 20, 27, 29, 30, 38, 40, 43, 45, 66, 69, 70, 75, 82, 89, 96, 97, 117, 118, 119, 120, 121, 122, 123, 125, 129, 132, 133, 135, 136, 139, 142, 149, 151
cost benefits, 66
counseling, 52
covering, 26, 137
CPI, 8
crabs, 131
crimes, 148, 164, 166, 169
criminal acts, 144
criminal punishment, ix, 159
criticism, 15
CRS report, 59
crude oil, 44, 59, 89, 119, 123, 131, 167
cure, 52
current limit, 123

D

danger, 147, 165
data analysis, 139
data collection, 91, 97, 105
data gathering, 84, 85
data set, 99
deaths, 160, 165, 167, 168, 171
decomposition, 109, 111
Deepwater Horizon incident, vii, 1, 2, 3, 10, 11, 12, 13, 14, 15, 33

defendants, 144, 146, 157, 159, 163, 164, 167, 168, 169, 171
degradation, 110, 111
degradation rate, 111
delinquency, 48, 49
Delta, 72
demonstrations, 12
Department of Commerce, 78, 79
Department of Energy, 113, 138
Department of Homeland Security, 26, 28, 124, 139
Department of Justice, 98, 112, 144, 164
Department of Labor, 29
Department of the Interior, 29, 78, 79, 84, 102, 113, 121, 138, 141
Department of Transportation, 25, 26, 28
Departments of Agriculture, 140
deposits, 104
depth, 33, 70, 84, 110
destruction, 6, 40, 41, 42, 77, 96, 151
deterrence, 148, 162
deviation, 164
DHS, 124
diesel fuel, 119, 123, 131
direct funds, 153
direct payment, 151
disaster, 10, 96, 111, 133
Disaster Relief Fund, 142
discharges, viii, ix, 35, 36, 39, 55, 77, 96, 139, 144, 145, 147, 149, 155
disposition, 49
dissolved oxygen, 111
distribution, 107
district courts, 57
District of Columbia, 51
diversification, 74
diversity, 105, 106, 107
DOI, 29, 121, 127, 135, 140
DOJ, 144, 145, 165, 171
DOT, 28
draft, 98
drinking water, 40, 42, 47, 77
dumping, 45

E

earnings, 19, 42
economic activity, 68
economic damage, vii, 1, 3, 8, 23, 25, 66, 76, 77, 83, 89, 96, 119, 122
economic damages, vii, 1, 3, 8, 23, 25, 66, 76, 77, 83, 89, 96
economic incentives, 73
economic losses, 6, 10, 16, 40
economic status, 120
economic theory, 66
ecosystem, 16, 98, 99, 100, 104, 114
effluent, 154
egg, 160
emergency, 10, 18, 70, 100, 101, 102, 103, 128
Emergency Economic Stabilization Act, 9, 14, 72, 128, 156
emergency response, 70
emission, 45
employees, 160, 161, 163, 166, 169
employment, 42, 68
endangered species, 159, 169
Endangered Species Act (ESA), ix, 159, 160
energy, 67
Energy Policy Act of 2005, 9, 14, 128
enforcement, 97, 128, 144, 146, 153, 154, 161
environment, 23, 27, 70, 86, 110, 118, 120, 122, 124, 131, 139
environmental harm, viii, 95
environmental impact, 86, 99, 112, 119, 123, 131
Environmental Protection Agency, 28, 57, 113, 145
environmental quality, 146
environmental regulations, 162
EPA, 15, 28, 29, 98, 127, 133, 145, 146, 147, 155
equipment, 45, 66, 70, 97, 103, 127, 129, 131, 132, 165, 166
erosion, 101, 104
evaporation, 131

evidence, 2, 8, 12, 18, 43, 46, 68, 109, 170, 172
Executive Order, 28, 29, 93, 141, 154
exercise, 29, 45, 59, 81
expenditures, 13, 125, 128, 138, 141, 149
expertise, 32, 69, 98, 103, 140, 169
exposure, 27, 71, 106, 107

F

farms, 104
fat, 55
federal agency, 18, 139
federal authorities, 10
federal environmental statutes, ix, 98, 144
federal government, viii, 2, 10, 11, 13, 29, 30, 36, 71, 75, 76, 78, 81, 86, 98, 114, 127, 149, 154
federal law, 19, 63, 141, 151, 153
Federal Register, 26, 28, 29, 84
federal regulations, 8, 140, 156
filtration, 98, 100
financial, viii, 2, 3, 8, 9, 12, 20, 27, 30, 46, 48, 52, 62, 63, 64, 65, 66, 67, 69, 70, 71, 72, 73, 86, 118, 122, 124, 125, 136, 139, 148, 168
financial condition, 136
financial incentives, 70
financial resources, 148
fish, viii, 23, 28, 40, 42, 47, 50, 75, 77, 81, 82, 84, 85, 91, 100, 106, 126, 131, 152, 166
Fish and Wildlife Service, 78, 79, 113, 114, 121, 133, 153, 154, 165
fisheries, 103, 110, 111
fishing, 76, 82, 84, 119, 122, 126, 131, 133, 135, 136, 142, 163
FMC, 165, 167, 171, 172
food, 170
force, 76, 91
foreclosure, 48, 49, 52
formation, 90
formula, 58
funding, 8, 9, 13, 15, 64, 71, 86, 97, 113, 149

funds, ix, 2, 10, 11, 12, 20, 21, 29, 32, 65, 70, 98, 114, 127, 128, 143, 144, 146, 149, 150, 152, 153, 154

G

GAO, 13, 30, 31, 72, 117, 118, 119, 126, 130, 139, 141
gas company, viii, 62
general taxpayers, vii, viii, 1, 2, 3, 20
George Mitchell, 30
God, 7, 43, 44, 59, 62, 151
government revenues, 41
governments, viii, ix, 39, 40, 75, 76, 78, 144, 150, 151, 161
governor, 10
grants, 128, 133, 152
grasses, 81, 84, 101
grounding, 135
grouping, 172
guardian, 94
guidance, 10, 19, 97, 103
guidelines, 63
guilty, 148, 156, 161, 164, 166, 167, 170, 171
Gulf Coast, 11, 15, 16, 19, 30, 31, 86, 94, 123, 133, 151
Gulf of Mexico, vii, viii, ix, 1, 7, 31, 32, 33, 35, 39, 57, 58, 59, 71, 73, 74, 75, 77, 90, 104, 110, 117, 118, 120, 121, 122, 128, 132, 133, 134, 137, 140, 141, 144, 155, 157, 159, 160

H

habitat, 81, 84, 85, 88, 101, 109, 114, 119, 122, 152, 154, 166, 169
habitats, 81, 84, 85, 88, 104, 105, 107, 109, 129, 140
harassment, 153, 157, 160
harvesting, 16
hazardous materials, 156
hazardous substances, 7, 28, 57, 97
hazards, 6, 42

health, 6, 42
heating oil, 131
heavy oil, 123, 131
historical data, 85, 99
history, ix, 10, 13, 31, 55, 70, 96, 146, 148, 164, 170
homeland security, 132
Homeland Security Act, 26
House, 11, 27, 29, 30, 31, 32, 56, 59, 72, 73, 74, 81, 88, 90, 94, 121
House Committee on Transportation and Infrastructure, 30, 121
House of Representatives, 27
human, 11, 63, 84, 86, 88
human health, 11, 63
hunting, 164
hurricanes, 82, 131, 142
hydrocarbons, 3, 110, 111

I

images, 109
impact assessment, 102
imprisonment, 147, 166, 167
improvements, 146
incarceration, 167
income, 42, 148, 149
Indians, 46, 92, 93
individual action, 161
individuals, 18, 64, 66, 75, 77, 87, 147, 162, 165
industries, 68
industry, 2, 11, 12, 19, 21, 67, 69, 70, 71, 104, 121, 122, 129, 132, 133, 139
inflation, 11, 31, 73, 118, 124, 127, 135, 142, 170
infrastructure, 70
ingestion, 123, 132
initiation, 128, 149
injure, 157, 160, 163
injuries, 19, 64, 75, 83, 84, 85, 151, 166
injury, 6, 19, 29, 40, 41, 42, 77, 80, 85, 88, 96, 98, 105, 147, 150
inspections, 52
institutions, 71, 102, 109, 139

internalization, 66, 73
invertebrates, 81
investment, 149
islands, 107, 161
issues, vii, 1, 2, 4, 11, 15, 20, 21, 30, 62, 72, 78, 80, 81, 82, 90, 91, 97, 113, 123, 160

J

jurisdiction, 47, 51, 57, 83, 98, 126, 160, 161

K

kill, ix, 153, 157, 159, 160, 163, 164, 165, 166, 169

L

larvae, 100, 106
law enforcement, 48, 49
laws, ix, 4, 10, 76, 83, 93, 135, 152, 157, 159, 160, 161, 169
lead, 68, 74, 80, 132, 159, 163
legislation, 9, 11, 15, 20, 64, 68, 72, 96, 137
legislative proposals, 27, 62
letters of credit, 63
liability insurance, 67
light, 43, 68, 105, 111, 123, 131, 133, 168
liquefied natural gas, 26, 28
litigation, 2, 7, 18, 27, 30, 67, 78, 80, 90, 112, 114
local government, ix, 40, 47, 48, 49, 78, 117, 120
logistics, 129, 140
Louisiana, 7, 17, 25, 26, 28, 73, 78, 79, 93, 102, 103, 113, 144, 161, 166

M

magnitude, 2, 11, 13, 21, 80, 105, 118, 122, 129, 137
majority, 102

mammal, ix, 107, 157, 159, 160, 163, 170
mammals, 81, 84, 103, 107, 109, 123, 132, 133, 152, 159, 160, 163
management, 6, 48, 49, 51, 52, 69, 71, 76, 81, 140, 172
mangroves, 81, 84, 109
manufacturing, 165
marine environment, 23
Marine Mammal Protection Act, ix, 76, 81, 144, 152, 154, 157, 159, 160, 163
Marine Mammal Protection Act (MMPA), ix, 159, 160
marsh, 82, 84, 88, 101, 107, 109
materials, 126
matter, 19, 83, 101, 112, 155
media, 30
methodology, 16, 32
Mexico, 32, 39, 120, 155, 160
migration, 157
Migratory Bird Treaty Act (MBTA), ix, 159, 160
migratory birds, ix, 78, 103, 153, 159, 165, 167, 170
mission, 132
Mississippi River, 104
moral hazard, 66
mortality, 107
multiplier, 109
mussels, 106

N

naming, 144
National Environmental Policy Act (NEPA), 86, 94
National Park Service, 78, 79
natural disaster, 44, 59
natural gas, 3
natural resource damages, vii, ix, 1, 3, 8, 10, 23, 25, 40, 63, 77, 89, 90, 96, 98, 100, 101, 144
natural resources, viii, ix, 6, 8, 19, 23, 24, 39, 40, 41, 42, 47, 62, 64, 66, 72, 75, 76, 77, 80, 81, 82, 83, 87, 89, 90, 91, 96, 97, 98, 99, 108, 109, 113, 124, 128, 133, 140, 144, 148, 150, 151
Natural Resources Damage Assessment (NRDA), viii, 75
non-government personnel, viii, 61, 95, 143
North America, 74, 148, 153, 154
nursing, 157
nutrient, 100, 104

O

Obama, 11, 16, 31, 73
Obama Administration, 11, 16, 31, 73
obstacles, 67
Office of Management and Budget, 14, 31
officials, 7, 67, 118, 121, 122, 128, 129, 131, 132, 133, 139
Oil Pollution Act of 1990, v, viii, ix, 4, 10, 21, 27, 28, 30, 35, 36, 37, 44, 53, 57, 62, 72, 73, 74, 77, 92, 96, 114, 117, 120, 144, 150, 151, 156, 157
Oil Pollution Act of 1990 (OPA), v, viii, ix, 4, 27, 28, 35, 36, 77, 92, 96, 117, 120
oil production, 71
oil spill costs, vii, 1, 3, 71, 117, 120, 121, 129
Oil Spill Liability Trust Fund, ix, 2, 4, 7, 9, 14, 29, 36, 37, 38, 40, 41, 42, 44, 46, 56, 57, 58, 63, 65, 70, 71, 76, 86, 89, 117, 119, 121, 130, 140, 144, 146, 147, 149, 154
omission, 7, 43, 62, 151, 168
operations, 2, 6, 11, 12, 49, 51, 68, 70, 139, 140, 163
opportunities, 76, 100
opt out, 74
organize, 91, 102
Outer Continental Shelf Lands Act, 5, 22, 23, 24, 26, 27, 46, 49, 50, 70, 113, 139, 140, 141, 161
overlap, 91
oversight, 112
ownership, 6, 23, 40, 44, 48, 49
ox, 111
oxygen, 104, 110, 111, 115

Index

oyster, 16, 81, 84
oysters, 88, 98, 106

P

parallel, 81, 88, 91, 114
penalties, vii, viii, ix, 1, 3, 14, 15, 29, 61, 63, 65, 66, 98, 114, 141, 143, 144, 145, 146, 147, 149, 151, 152, 153, 154, 155, 159, 167, 168, 171
permission, 88
permit, 5, 49, 50, 162, 163
personal communication, 14
pesticide, 165
petroleum, 44, 47, 56, 110, 118, 120, 122, 128, 131, 141, 149
Petroleum, 27, 74, 92, 93, 171
plankton, 106
plants, 152
platform, 33, 133
polar, 153, 154
policy, viii, 3, 11, 62, 64, 68, 69, 95, 146
policy issues, 62
policy options, 62
policy reform, 68
policymakers, vii, 1, 2, 4, 11, 20
pollutants, 7
pollution, 28, 50, 57, 98, 140, 150
polycyclic aromatic hydrocarbon, 111
ponds, 165
population, 100
precedent, 166
preparation, 84, 114
President, 4, 7, 8, 10, 25, 26, 27, 28, 30, 32, 37, 44, 55, 56, 70, 73, 74, 93, 113, 128, 135, 141
Prince William Sound, 3, 91, 150
private firms, 71
private party, 3, 6
probability, 33, 67
profit, 6, 41
project, 99, 103, 114
propane, 110, 111
proposition, 68
protection, 6, 42, 76, 146, 153, 154

public concern, 139
public health, 28, 50, 139, 146
public interest, 89, 129, 132
public service, 6, 23, 42, 43, 72, 151
Puerto Rico, 51
punishment, ix, 147, 148, 159, 165

Q

quantification, 85, 99, 100

R

radio, 107
ramp, 108
real property, 6, 41, 42
real time, 113
reasoning, 12, 166
recognition, 148
recommendations, 33, 73, 118, 121, 124, 127, 134, 135
recovery, 14, 16, 29, 30, 39, 58, 76, 77, 83, 88, 89, 90, 97, 100, 118, 119, 122, 125, 132, 144, 149, 150, 153, 154
recovery plan, 76
recreation, 82, 84
recreational, 82, 84, 100, 123, 133
Reform, 55
reforms, 69
regulations, 3, 6, 7, 8, 9, 19, 25, 26, 36, 39, 40, 56, 57, 59, 63, 78, 80, 82, 83, 85, 93, 94, 96, 97, 99, 104, 109, 114, 122, 124, 141, 164
regulatory agencies, 139
regulatory requirements, 49, 83
rehabilitation, 80
reimburse, 76, 82, 114, 151
reinsurance, 67
relief, 58, 133
remediation, 85, 86
removal actions, 37
repair, 148, 150
requirements, 10, 12, 37, 40, 59, 64, 65, 68, 69, 73, 83, 94, 157, 162

researchers, 91
reserves, 70
resolution, 90, 91, 97
resources, viii, ix, 6, 20, 21, 23, 40, 41, 42, 47, 62, 70, 75, 76, 77, 78, 80, 81, 82, 83, 84, 85, 86, 89, 90, 91, 96, 97, 98, 100, 102, 104, 106, 109, 114, 123, 137, 140, 150, 151, 154, 155, 161
response, 2, 3, 6, 7, 8, 11, 12, 13, 16, 21, 28, 29, 30, 55, 63, 64, 72, 76, 77, 82, 84, 85, 89, 98, 100, 102, 109, 113, 114, 119, 120, 121, 122, 125, 127, 128, 129, 131, 132, 133, 137, 138, 139, 140, 142, 150, 151, 156, 157
restitution, 76, 148, 149, 154, 166
restoration, viii, 2, 15, 20, 24, 75, 76, 77, 78, 80, 81, 83, 84, 85, 86, 88, 89, 90, 91, 95, 96, 98, 99, 100, 101, 103, 104, 105, 109, 112, 113, 114, 128, 144, 149, 150, 151, 154
restructuring, 52
revenue, 8, 14, 15, 23, 29, 41, 123, 128, 131, 133, 138, 141, 168
risk, 12, 20, 21, 56, 64, 65, 66, 67, 69, 70, 73, 74, 81, 104, 119, 123, 133, 139
risk assessment, 20, 139
risk factors, 20
risk management, 69, 70, 74
risks, 20, 55, 63, 67, 71, 74, 119, 120, 123, 128, 134, 137
risk-taking, 71
rule of law, 37, 38
rules, 160
runoff, 104, 167

S

safety, 6, 7, 39, 42, 54, 62, 63, 65, 66, 69, 70, 122, 124, 134, 151
salmon, 112
scaling, 99
scatter, 132
scientific knowledge, 97
scientific method, 97

scope, 2, 6, 11, 12, 31, 32, 59, 66, 85, 152, 169
sea level, 104
Secretary of Homeland Security, 141
security, 48, 49, 51, 52, 53
sediment, 81, 84, 103, 104, 109
seizure, 48, 49
Senate, 11, 18, 27, 30, 31, 32, 56, 59, 68, 70, 93, 121, 164
sensitivity, 68
sentencing, 171
Sentencing Guidelines, 148
services, 6, 42, 46, 82, 85, 98, 99, 100, 114, 129
settlements, 146
shellfish, 28, 50, 81, 84, 106
shoot, 157, 164, 169
shoreline, 81, 82, 84, 105, 109, 112, 119, 122, 129, 131
shores, viii, 75, 76, 85
shrimp, 106
signals, 109
sludge, 47
small firms, 74
specialists, 112
species, ix, 76, 81, 83, 85, 107, 109, 133, 140, 152, 154, 159, 160, 162, 169
speech, 154
stabilization, 104
Staff Working Papers, viii, 95
staffing, 102
Stafford Act, 10, 30, 142
stakeholders, 18
state, ix, 10, 16, 17, 37, 40, 42, 43, 44, 49, 57, 63, 65, 66, 75, 76, 78, 80, 81, 84, 102, 103, 104, 114, 117, 120, 121, 127, 129, 131, 144, 150, 151, 154, 156, 161, 169, 171, 172
state control, 81
state laws, 10, 144
states, viii, 5, 10, 16, 19, 23, 24, 25, 31, 37, 38, 39, 40, 41, 42, 66, 67, 68, 75, 76, 78, 80, 81, 82, 83, 85, 91, 97, 98, 113, 114, 123, 133, 139, 140, 147, 152, 161

statutes, ix, 5, 7, 26, 77, 98, 144, 146, 149, 160, 162, 168
statutory provisions, 15
steel, 167
storage, 55, 150
storms, 131
structure, 8, 20, 45
subsidy, 12
subsistence, 6, 23, 41, 72, 76, 82, 87, 150
Supreme Court, 27, 90, 170, 171

T

tanks, 126, 150, 165
tar, 109, 132
target, 105
tax increase, 141
taxes, 6, 41
taxpayers, vii, viii, 1, 2, 3, 20
technical support, 102
techniques, 71
technology, 107
territorial, 45, 47, 51, 56, 81, 93, 160
territory, 51, 81
testing, 101
The Homeland Security Act, 28
threats, 135
tides, 132
tissue, 106, 109
Title V, 76, 90
total costs, viii, 3, 42, 61, 120, 121, 123, 133, 139
tourism, 123, 131, 133
toxicity, 102, 110, 111
training, 101, 114
transparency, 16
transport, 135
transportation, 28, 51, 55, 56
Treasury, 44, 59
trial, 83, 90, 165
tribal lands, 78
trust fund, 2, 4, 9, 10, 13, 14, 20, 21, 29, 30, 31, 44, 56, 140
Trust Fund, vi, ix, 2, 4, 7, 9, 14, 29, 36, 37, 38, 40, 41, 42, 44, 46, 56, 57, 58, 63, 64, 65, 70, 71, 76, 86, 89, 94, 117, 119, 121, 130, 140, 144, 146, 147, 149, 150, 154
turtle, 169

U

U.S. history, 137, 160
U.S. Treasury, 15, 74, 149, 153
underwriting, 67
uniform, 167
universities, 138
unlimited liability, 7, 12, 66, 74, 125, 152

V

vegetable oil, 55
vegetation, 81, 101, 103, 105
vehicles, 27, 47
vessels, 8, 9, 20, 22, 23, 24, 25, 26, 28, 29, 39, 46, 55, 72, 96, 100, 104, 109, 118, 119, 120, 121, 123, 124, 125, 126, 127, 135, 136, 137, 139, 142
victims, 64, 65, 67, 71, 148
Viking, 94
Vitter, 32
vote, 11, 80
vulnerability, 13, 30

W

war, 7, 43, 62, 80, 103, 151
Washington, 71, 138, 139, 141, 142
waste, 101
wastewater, 165
water, 28, 40, 42, 47, 50, 51, 54, 70, 77, 81, 84, 98, 103, 104, 105, 106, 109, 110, 111, 115, 125, 126, 127, 128, 129, 131, 167
watershed, 104
waterways, 126
welfare, 28, 50, 139
wells, 33
wetlands, 103, 104, 153, 154
whales, 112, 114, 163

White House, 32
wildlife, viii, ix, 23, 28, 40, 42, 47, 50, 75, 76, 77, 81, 84, 85, 93, 109, 111, 112, 133, 144, 152, 153, 157, 159, 160, 166, 168, 169, 171
workers, 132

working groups, 102

Y

yield, 2, 21, 145